Environmental Quality, Monitoring and Management

Environmental Quality, Monitoring and Management

Editor: Meghan Higgins

Callisto Reference

www.callistoreference.com

Callisto Reference,
118-35 Queens Blvd., Suite 400,
Forest Hills, NY 11375, USA

Visit us on the World Wide Web at:
www.callistoreference.com

ISBN: 978-1-64116-218-0 (Hardback)

Printed in China.

Cataloging-in-Publication Data

Environmental quality, monitoring and management / edited by Meghan Higgins.
p. cm.
Includes bibliographical references and index.
ISBN 978-1-64116-218-0
1. Environmental quality. 2. Environmental monitoring. 3. Environmental management.
4. Environmental protection. I. Higgins, Meghan.

GE140 .E58 2019
363.7--dc23

Table of Contents

Preface

Over the recent decade, advancements and applications have progressed exponentially. This has led to the increased interest in this field and projects are being conducted to enhance knowledge. The main objective of this book is to present some of the critical challenges and provide insights into possible solutions. This book will answer the varied questions that arise in the field and also provide an increased scope for furthering studies.

Environmental quality refers to the measure of the environmental condition relative to human need and the requirements of one or more species. It is understood with respect to the natural and the built environment, including water pollution, air purity, noise, etc. and the effects of such characteristics on physical and psychological health. The monitoring of environmental quality is useful in environmental impact assessment, as well as in the circumstances when anthropogenic activities pose an environmental risk. Environmental monitoring is done with respect to air, soil and water quality monitoring, which is aided by various sampling methods. Environmental resource management is concerned with the management of the impact of human activities on the environment. It involves the management of both the biotic and abiotic components of the biophysical environment and the relationships among living organisms and their habitats. This book provides significant information of environmental quality, monitoring and management to help develop a better understanding of the current environmental condition and its management. It is a collective contribution of a renowned group of international experts. This book is a vital tool for all researching or studying these fields as it gives incredible insights into emerging trends and concepts.

I hope that this book, with its visionary approach, will be a valuable addition and will promote interest among readers. Each of the authors has provided their extraordinary competence in their specific fields by providing different perspectives as they come from diverse nations and regions. I thank them for their contributions.

Editor

1

Biomarkers and Ultra Structural Evaluation of Marine Pollution by Polycyclic Aromatic Hydrocarbons

Nehad M. Abd El Moneam[1], Nabila E. Abd El Maguid[1], Amany M. El-Sikaily[2], Marwa G. Zaki[2], Mohamed A. Sheradah[3*]

[1]Biochemistry and Zoology Department, Faculty of Science, Alexandria University, Alexandria, Egypt
[2]Marine Pollution Research Lab., National Institute of Oceanography & Fisheries, Alexandria, Egypt
[3]Marine Biotechnology Lab., National Institute of Oceanography & Fisheries, Alexandria, Egypt
Email: *niof.shreadah@gmail.com

Abstract

The present study is endeavored to study the oxidative stress and antioxidant response as well as the ultrastructural change of both liver and muscle tissues of *Siganus rivulatus* collected from polluted region (El-Mex Bay) and control region (Marsa Matrouh) to identify the significance of biomarkers. The results revealed that, El-Mex Bay was polluted by different types of chemicals including hydrocarbons comparing with the reference area (Matrouh area). Where, the concentration range of total hydrocarbons in sediment samples collected from Matrouh coast was 409.24 - 521.26 ng/g, dry weight, where as this range in sediment samples collected from El-MexBay was 4159.77 - 4589.81 ng/g, dry weight. In this context, the induction of antioxidant systems in fish collected from polluted area in response to oxidative stress should be considered as a clear indication of the presence of pollution and environmental health degradation. Also, The increase in lipid peroxidation was a useful indicator of the pollution load in the present study. The results suggest that chemical pollution is capable of inducing morphological alteration in liver offish collected from polluted area. The present study indicated that ultra-structural changes serveas biomarker of stress in aquatic environment.

Keywords

Biomarkers, Ultra Structural Evaluation, Polycyclic Aromatic Hydrocarbons (PAHs)

1. Introduction

Environmental pollution has increased substantially in the last decades due to a great number of industrial, agricultural, commercial and domestic waste, effluents and emissions as well as hazardous substances. Aquatic pollution is the predominant form of pollution since the majority of chemical pollutants are entering to sea, rivers, lakes and wetlands [1]-[9]. Marine and coastal ecosystems are characterized by their complexity and their sensitivity to various inorganic and organic pollutants [10]-[18]. Because environmental contaminants can have a broad spectrum of sub lethal effects on organisms, so bioindicators are useful tools for assessing the presence and levels of chemical pollution. Such effects in organisms sensitive to contaminant exposures can be used as early warning signs for the degradation of the environment [19]. Polycyclic aromatic hydrocarbons (PAHs) are a class of organic compounds consisting of two or more fused aromatic rings. They are one of the major categories of pollutants entering the marine environment and finally accumulating in the sediments. Polycyclic aromatic hydrocarbons (PAHs) can accumulate in the tissues of aquatic animals and as such tissue concentrations of chemical pollutants can be of public health concern to both animals and humans [20]-[27]. They are formed by three main processes: diagenesis, petrogenesis, and pyrolysis [28]. Diagenesis and petrogenesis are typically naturally-occurring, where as pyrolysis can be the result of natural or anthropogenic events [29]. Diagenesis occurs in organic matter that has been deposited in soils or sediments [30]. Pollutants accumulated in tissues of fish may catalyze reactions that generate reactive oxygen species (ROS) which may lead to environmental oxidative stress. Indicators of oxidative stress include changes in antioxidant enzyme activity, damaged DNA bases, protein oxidation products, and lipid peroxidation products [31]. Disturbance of living processes at the molecular and subcellular levels of biological organization by xenobiotics can lead to cell injury, resulting in degenerative and neoplastic diseases in target organs. Therefore histopathological biomarkers have been proven to be useful indicators of toxicity in fish organs [32].

The removal of xenobiotics, and even some endogenous substances, from the cell is catalyzed by a number of different enzymes, so called phase I and II enzymes. Phase I enzymes are involved in xenobiotic biotransformation via the introduction of a polar moiety which rendersa lipophilic contaminant to more hydrophilic. The cytochrome P450 (CYP) family is the most well studied in fish, especially CYP1A. CYP1A serves to increase the solubility of hydrophobic molecules through a reduction reaction involving an oxygen molecule [33]. Examples of phase II enzymes commonly used in biomonitoring programs involving fish are glutathione S-transferase (GST) and UDP glucuronyl transferase (UDPGT) [34]. The activity of phase I enzymes can lead to an increase in ROS production or the generation of reactive redox cycling intermediates. Antioxidant enzymes facilitate the removal of these reactive chemical intermediates and resulting ROS molecules. The action ofCYP1A can result in the production of $O_2^{\bullet}$—which in turn can be metabolized by superoxide dismutase (SOD) to H_2O_2. This hydrogen peroxide molecule can then be reduced to H_2O and O_2 by catalase (CAT) [35]. Hydroxyl radicals ($OH^{\bullet}$) can form both H_2O_2 and $O_2^{\bullet}$—via reactions with redox cycling metal ions, for example iron and copper. This highly potent hydroxyl radical can attack both protein and lipid molecules to form oxidative damage products.

Lipid damage can occur as a result of oxidative stress or a disruption in the balance between prooxidant and antioxidant factors. Reactive oxygen species are known to extract hydrogen atoms from unsaturated bonds thereby altering lipid structure or function [36].

Since fish species, *Siganus rivulatus*, is considered as a sentinel species for environmental monitoring, it was chosen to study the oxidative stress and antioxidant response as well as the ultra-structural change of both liver and muscle tissues of *Siganus rivulatus* collected from polluted region, *i.e.* El-Mex Bay and control region, *i.e.* Marsa Matrouh to identify the significance of biomarkers.

2. Materials and Methods

2.1. Study Area

Matrouh is located in the north-western Mediterranean coast of Egypt 290 km west of Alexandria. It is a protected area far from land-based sources of pollution. On the other hand, El-Mex Bay located west of Alexandria, is a semi-elliptical open basin, extends for about 15km between El-agamy head land in the west and the western harbor to the east and from the coast to a depth line of about 15 km. the bay has a mean depth of about 10 m and surface area of about 19.4 km^2. El-Mex Bay is a highly polluted area, the major types of pollution sources are domestic sewage, industrial waste water, and agricultural run-off, through lake out lets, and river discharged and oil pollution. El-Mex Bay receives mixed agricultural run-off from lake Mariout through El-max pumping station and El-umumdrain, industrial water from chloro-alkali plant, tanneries and slaughterhouse, also, air borne particles from the fumes of adjacent industrial plants including a cement factory [5] [11] [37]-[46]. Petroleum product from Al-Alamien oil field and from Suez Mediterranean pipeline terminal (SUMED) also contaminate the bay [47].

2.2. Sampling

Sediment samples were collected during March 2009 from ten stations, distributed along El-Mex Bay of Alexandria and, ten stations distributed along Matrouh coast as shown in **Figure 1(a)** & **Figure 1(b)**, respectively. At the same time Fish species (*Siganus rivulatus*), which belong to "*Siganidea*" were collected from El-Mex Bay and also from Matrouh coast. About 100 fish of similar weight (30 - 40 g) were collected from each area. Samples were dissected and their liver as well as, muscles was removed and prepared for electron microscopic studies. While the remaining liver and muscle tissue kept frozen at −20˚C until use. Fish liver samples were weighted to the nearest milligrams and their hepato-somatic index (HIS, liver weight/bodyweight × 100) was calculated according to Jangaard *et al.* [48].

2.3. Methodology

2.3.1. Morphological Study

Morphological features of *Siganus rivulatus* collected from both areas, *i.e.* El-Mex Bay

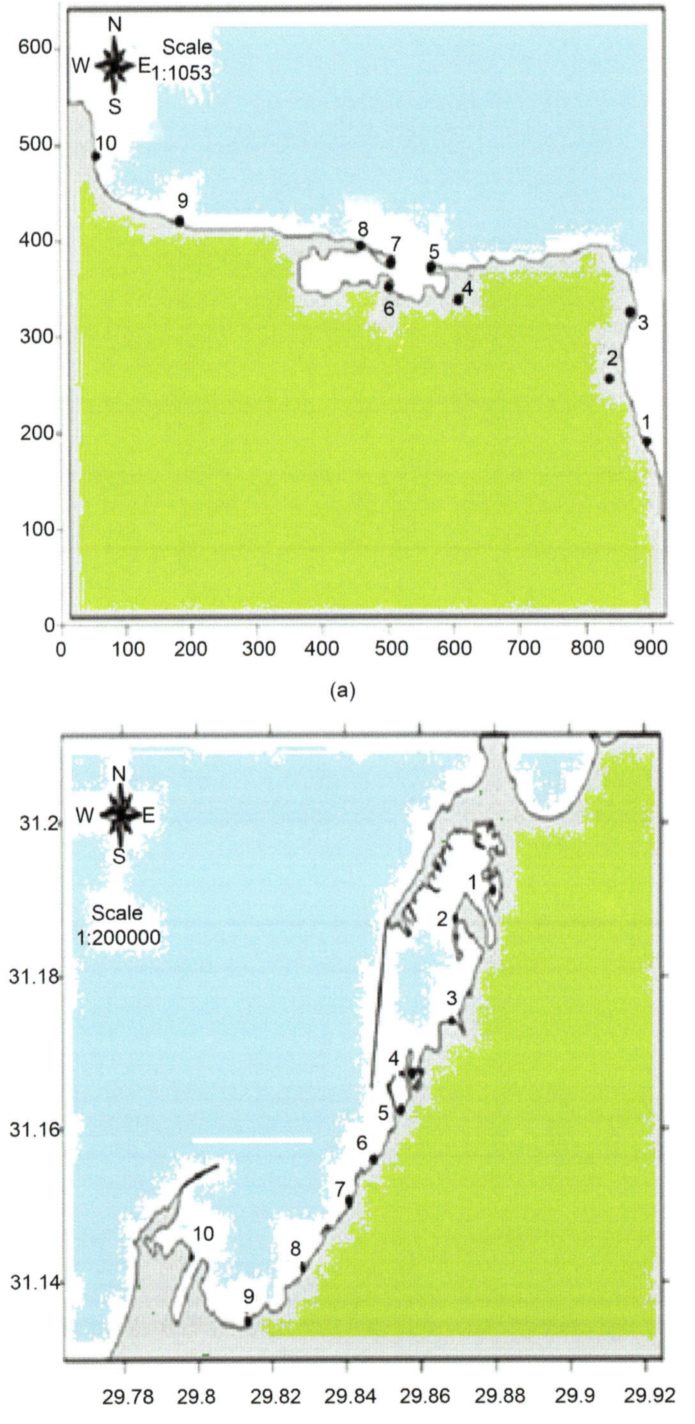

Figure 1. Location of sampling stations in the study area (a) Matrouh area and (b) El-Mex Bey area.

and Matrouh coast were determined. Both weight and length of *Siganus rivulatus* collected from both areas were taken and the hepato-somatic index (HIS = liver weight/body weight × 100) of fish was calculated.

2.3.2. Chemical Analysis

Hydrocarbons have been determined in sediment and fish according to UNEP/IOC/IAEA [49]. The samples were analyzed for aliphatic and aromatic hydrocarbons following these steps extraction, cleaning up and fractionation as well as their instrumental analysis followed by their analytical quality control. To control analytical reliability and assurance recovery efficiency as well as accuracy of the results, six analyses were conducted using hydrocarbon reference materials, HS-5 (sediment) provided by NRC-IMB of Canada and SRM 2974 (freeze-dried mussel tissue) (*Mytilus edulis*) provided by NIST of USA As well as sediment samples of known hydrocarbon levels spiked with a mixture consisting of 2.0 μg from each hydrocarbons were analyzed as above to validate the analytical method used in this study. The laboratory results showed recovery efficiency ranged from 90% - 110% for HS-5, 85% - 97% for SRM-2974 and 94% - 102% for the spiked samples. All solvents were pesticide grade.

2.3.3. Biochemical Analysis

1) Preparation of tissue homogenate

Prior to dissection, liver tissue was perfused with a (phosphate buffer saline solution, pH 7.4 (PBS), containing 0.16 mg/ml heparin to remove any red blood cells. The tissue was homogenized in 5 - 10 ml cold buffer (100 mM potassium phosphate, pH 7.0, containing 2 m MEDTA)/g tissue, and then centrifuge at 10,000 ×g for 15 min. at 4˚C. The supernatant was removed for assay. The biodiagnostic Glutathione *S*-transferase assay kit (CAT. No. GT 25 19) is used to measures total Glutathione *S*-Transferase activity by measuring the conjugation of 1-chloro-2, 4-dinitrobenzene (CDNB) with reduced glutathione. The conjugation is accompanied by an increase in absorbance at 340 nm [50]. Reduced Glutathione (GSH) concentration is determined by Biodiagnostic Gltathione reduced kit (CAT. No. GR 2511). The method based on the reduction of 5, 5' dithiobis (2-nitrobenzoic acid) (DTNB) with glutathione (GSH) to produce a yellow compound, the reduced chromogen directly proportional to GSH concentration and its absorbance can be measured at 405 nm [51]. The determination of lipid peroxidation is based on the reaction of malondialdehyde (break down product of lipid peroxides) with thiobarbituric acid (TBA), the resulting thiobarbituric acid reacting substances (TBARS) concentration, was determined by measuring the absorbance at 535 nm [52].

2) Microscopic study

Ultra thin sections (50 nm) were cut using LKB ultra tome with a glass knife, mounted on copper grids. Semi thin sections (1 μm) were also cut from the sample blocks, stained with toluidine blue, and examined under a light microscope. Ultra thin sections were doublestained with freshly prepared uranyl acetate [53] for 20 minutes as well as leadcitrate for 5 minutes, and scoping grids was achieved by using JEOL 100CX transmission electron microscope of the Faculty of Science Alexandria University.

3) Statistical analysis:

This study was statistically analyzed using SPSS (version 10). A difference is considered as significant at $p < 0.05$.

3. Results and Discussion

3.1. Mophological Study

The morphological study showed that the weight and length range for *Siganus rivulatus* collected from each area were relatively similar, but the hepatosomatic index for *siganus rivulatus* collected from El-Mex Bay was 2 fold more than those collected from Matrouh coast (**Figure 2**). Statistical analysis of data revealed a highly significant difference between the two investigated areas, *i.e.* El-Mex area and Matrouh area at $p < 0.05$.

3.2. Chemical Analysis

The results of the chemical analysis revealed that the concentration range of total hydrocarbons in sediment samples collected from Matrouh coast was 409.24 - 521.26 ng/g, dry weight, whereas the range in sediment samples collected from El-Mex Bay was 4159.77 - 4589.81 ng/g, dry weight revealing the presence of high concentrations of total hydrocarbons in El-Mex Bay (tested area) compared to the reference area (Matrouh area). The statistical analysis of total hydrocarbons in sediment samples revealed a highly significant difference between the two areas at $p < 0.05$. The concentration of total hydrocarbons in sediment collected from El-Mex Bay (tested area) was 9.5 fold more than those of Matrouh coast (control area) as shown in **Figure 3**. The highest concentration was observed in sediments collected from station 7 (4589.81 ng/g dry weight) followed by stations 9, 1, 5, while the lower concentrations were measured in samples collected from station 6. These would suggested that hydrocarbons accumulated in Mediterranean sea sediments come from nearly human activities and fuel combustion emissions [5] [17] [25] [26] Kim *et al.* [54] demonstrated that the nature of the sediment influences the distribution and concentration of PAHs. Concentrations of PAHs in sediment were affected by chemical composition of the sediment such as organic matter and clay content. Sediments with high organic carbon content was characterized

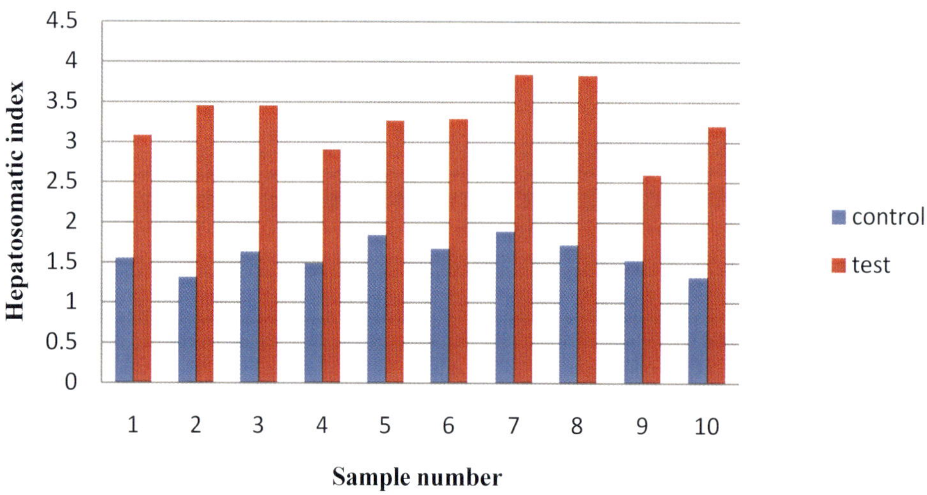

Figure 2. Hepatosomatic index of "*Siganus rivulatus*" collected from El-Mex Bay compared to Matrouh coast.

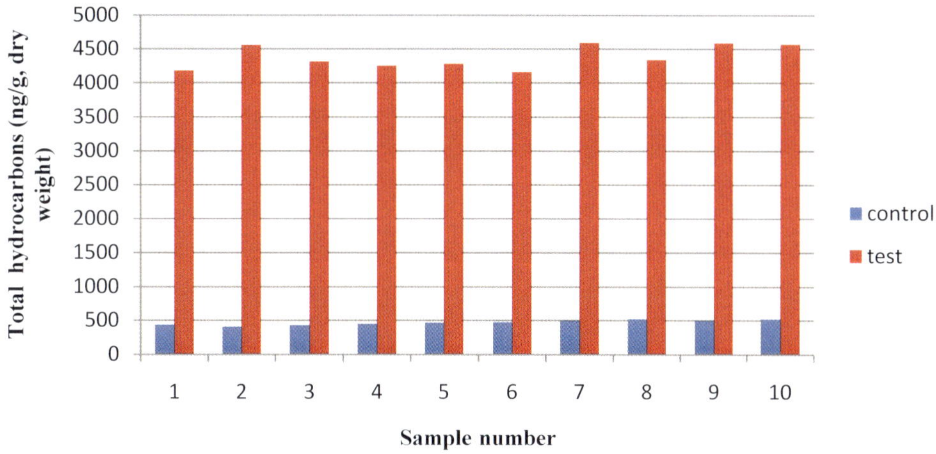

Figure 3. Total hydrocarbons concentration (ng/g, dry weight) of sediment samples Collected from El-Mex Bay compared to Matrouh coast.

by its high values of PAH's [55] [56]. Moreover, Simpson *et al.,* [57] showed that the relationship between total PAHs in sediments and organic carbon was only significant for highly contaminated sites, containing PAHs concentration greater than 2000 ng/g dry weight. In the present study, all sediment samples collected from El-Mex Bay were characterized by its higher total organic carbon (TOC). The importance of sedimentary organic matter on the partitioning of PAHs in sediments has been well documented [58]. Leung *et al.* [19] found that the high partitioning of PAHs to sedimentary organic matter was mainly due to the significant aromatic fraction of the organic matter.

To determine the source of PAHs, the ratios of phenanthrene/anthracene, fluoranthrene/pyrene and Fluoranthene/Fluoranthene + Pyrene were used [59]-[61]. Where, phe/anth more than 10 indicated to petrogenic input. The ratio of flu/pyr less than 1 was attributed to petrogenic sources and when it is greater than 1 was obviously related to a pyrolytic origin. The present study revealed that, the ratio of phe/anth was less than 10 in both tested and control areas, suggesting that their PAH were pyrolytic-derived. All samples from the control area gave also flu/pyr ratio greater than 1 indicating to a pyrolytic origin of PAHs in this area. On contrast, sediments of El-Mex Bay area gave flu/pyr ratio less than 1 indicating a petrogenic origin of PAHs.

In *Siganus rivulatus* muscles the concentration range of total hydrocarbons collected from Matrouh coast was 104.72 - 219.18 ng/g, dry weight, whereas, this range in *Siganus rivulatus* muscles collected from El-Mex Bay was 2239.52 - 3532.11 ng/g, dry weight. There was a very high significantly difference between the two areas at $p < 0.05$. The concentration of total hydrocarbons in fish muscles collected from El-Mex Bay (tested area) was 16 fold more than those collected from Matrouh coast (control area) as shown in **Figure 4**. An elevation of PAHs in fish samples but it is still less than that found in sediments and higher than that present in the water column due to the increasing metabolism [25]-[27]. PAHs in the fish were mostly dominated by the high molecular weight PAHs (4 - 6 rings). Several studies [62] [63] have shown that fish preferentially bioaccumulate 4, 5 and 6 ringed PAHs rather than 2, 3 ringed PAHs. In

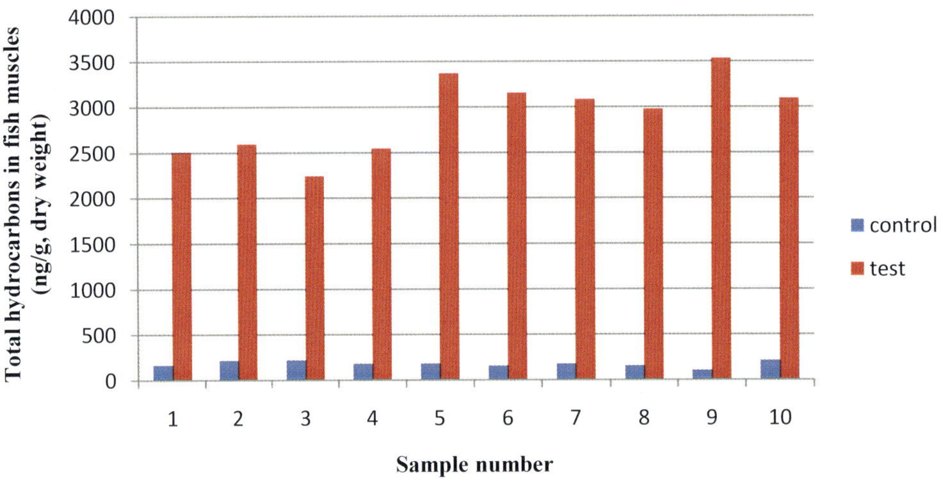

Figure 4. Total hydrocarbons concentration (ng/g, dry weight) of fish muscles collected from El-Mex Bay) compared to Matrouh coast.

the present study, some samples of Matrouh had pr/ph ratio greater than 1 and other samples have undetected phytane this indicated that most PAHs of this control area get from zooplankton, while El-Mex Bay showed that pristine and phytane were present in all samples indicating the petroleum origin of PAHs in this area.

3.3. Biochemical Analysis

Contamination of water with industrial and agricultural pollutants influences the biochemical processes of aquatic organisms [64]. An effective monitoring system using biochemical markers has been established to demonstrate these xenobiotics in the environment [65]. On the other hand, the evaluation of oxidative stress is commonly used in monitoring programs based on measurements of catalyze (CAT), glutathione peroxidase (GPX) and superoxide dismutase (SOD) activities [66]-[68]. The generation of oxyradicals has a connection to the etiology of several human diseases and is probably the same for fish [69]. The statistical analysis of liver and muscle glutathione s-transferase revealed that there was a highly significant increase in liver and muscle glutathione s-transferase activity and glutathione concentration of *Siganus rivulatus* collected from El-Mex Bay than those collected from Matrouh coast at $p < 0.05$. Liver and muscle glutathione s-transferase activity of *Siganus rivulatus* collected from El-Mex Bay increased by 107.21% and 63.41%, respectively, compared to those collected from Matrouh coast as shown in **Figure 5** and **Figure 6**, respectively. Apart from the normal metabolism in a living organism; carcinogens (pesticides, heavy metals), infections (bacterial, parasitical, viral), radiation damage, and environmental stress factors cause an increase in free oxygen radicals and thus cause oxidative stress [70]. Recent findings show also that the pollution toxicity in an aquatic organism may be connected to increased production of reactive oxygen species (ROS) that leads to oxidative stress [71] [72]. In this respect, a number of studies confirmed the successful role of antioxidant enzymes and non-enzymatic antioxidant modulation in identifying environmental stress [73]. In addition,

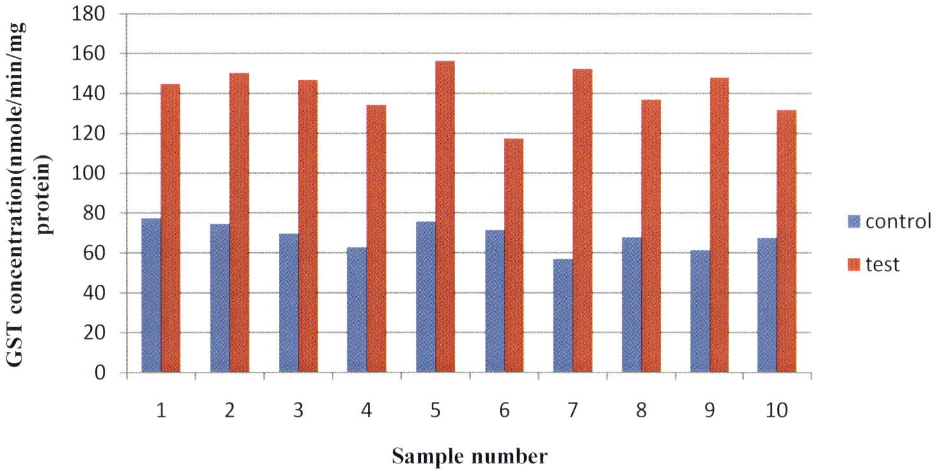

Figure 5. Liver glutathione s-transferase activity of "*Siganus rivulatus*" collected from El-Mex Bay compared to Matrouh coast.

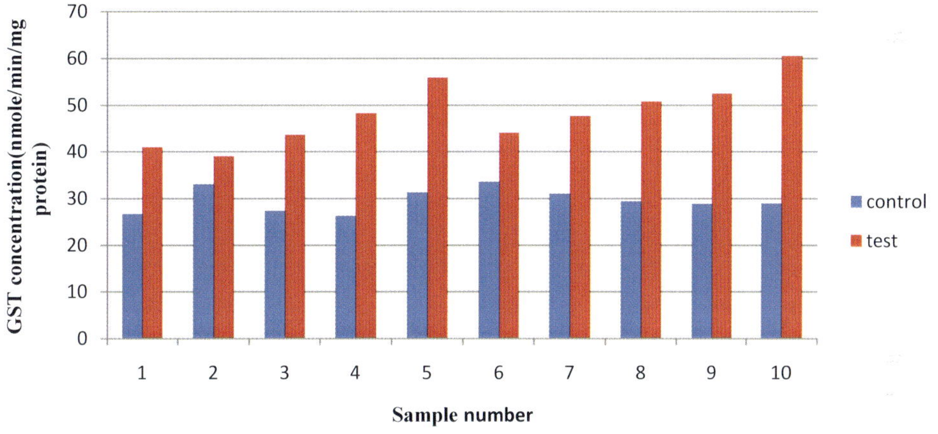

Figure 6. Muscle glutathione s-transferase activity of "*Siganus rivulatus*" collected from El-Mex Bay compared to Matrouh coast.

lipid peroxidation estimation in particular has also been found to have a high predictive importance as a biomarker of exposure [74]-[76]. Therefore, this study aimed to detect the oxidation stress response in liver and muscle of *Siganus rivulatus* collected from El-Mex Bay (tested area) and Matrouh (controlarea) to identify the significance of oxidative biomarkers as an early indicator of the health of marine ecosystem. Antioxidant enzymes were included in this study because of their inducibility under conditions of oxidative stress and their potential role in adaptation to pollutant-induced stress. An important liver biomarkers used for environmental pollution are glutathione S-transferase (GST) enzyme and glutathione in its reduced form (GSH) [77]. The presence of organic pollutants in fish muscle and their relation to specific biomarkers were also determined by Siroka *et al.,* [78]. Where, the GST is an important intracellular enzyme of the second stage of xenobiotic metabolism. Its main function is to catalyze the conjugation of glutathione and electrophilic substances of exogenous origin that might be one of etiological factors of carcinogenesis and development of degenerative diseases [79]. It is used as a biochemical marker of aquatic environmental contamination with exogenous substances [80].

The liver and muscle tripeptide glutathione concentration of *Siganus rivulatus* collected from El-Mex Bay was increased by 73.63% and 77.54%, respectively compared to its corresponding value collected from Matrouh coast (**Figure 7** and **Figure 8**, respectively). The liver and muscle lipid peroxidation concentration for "*Siganus rivulatus*" collected from El-Mex Bay increased by 255% and 176%, respectively, compared to those collected from Matrouh coast (**Figure 9** and **Figure 10**, respectively). Moreover, measurements of lipid peroxidation which has been described as a biomarker for effect of pollution in several studies was also a useful indicator for the pollution load [71]. An elevation of MDA registered the oxidative stress due to the presence of contamination [79]. This study indicated the existence of a significant elevation in liver and muscle lipid peroxidation of "*Siganus rivulatus*" collected from El-Mex Bay compared to that

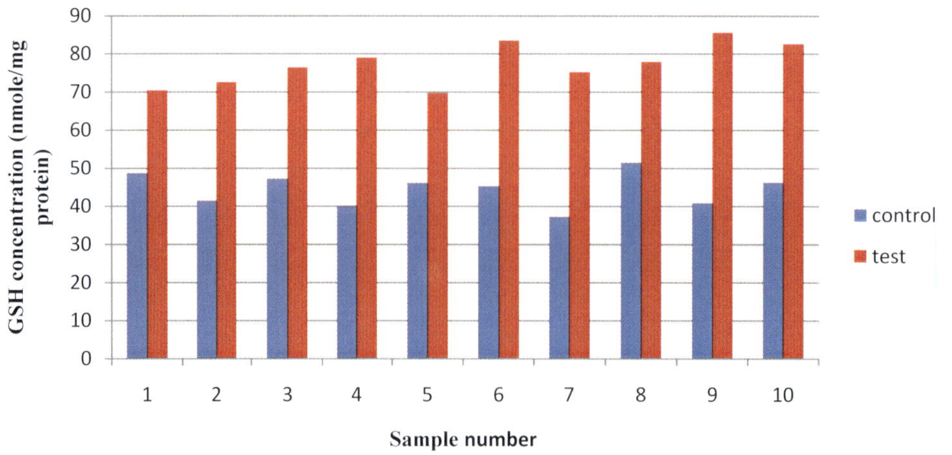

Figure 7. Liver glutathione reduced concentration of "*Siganus rivulatus*" collected from El-Mex Bay compared to Matrouh coast.

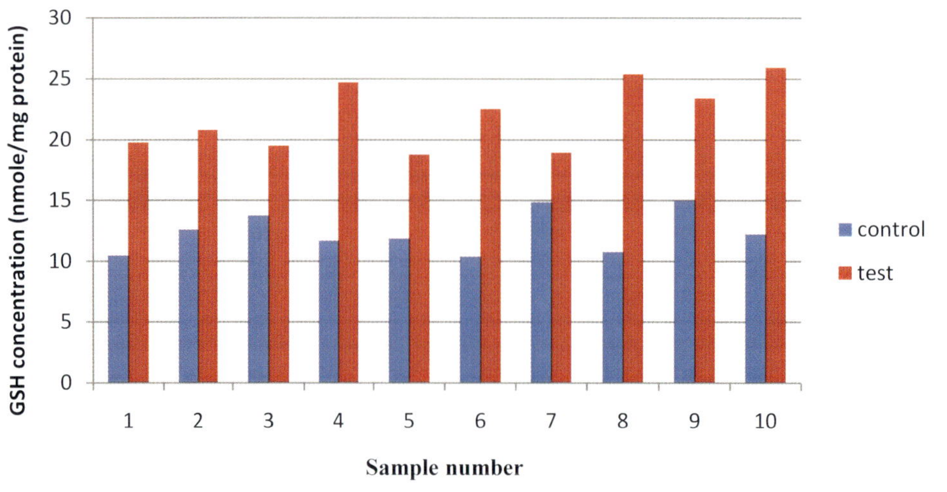

Figure 8. Muscle glutathione reduced concentration of "*Siganus rivulatus*" collected from El-Mex Bay compared to Matrouh coast.

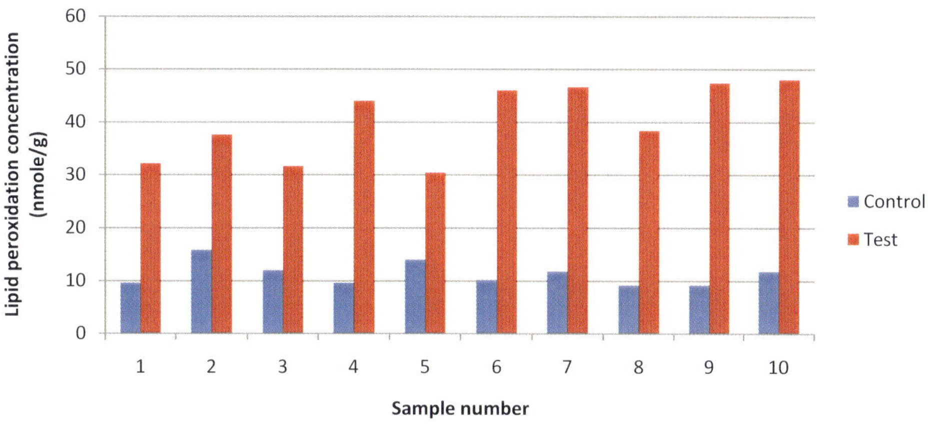

Figure 9. Liver lipid peroxidation concentration (nmole/g) of "*Siganus rivulatus*" collected from El-Mex Bay compared to Matrouh coast.

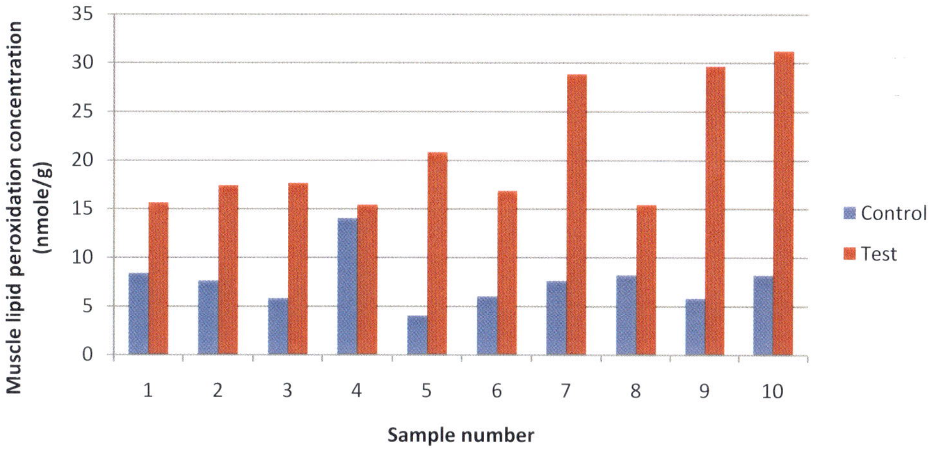

Figure 10. Muscle lipid peroxidation concentration (nmole/g) of "*Siganus rivulatus*" collected from El-Mex Bay compared to Matrouh coast.

collected from Matrouhcoast (Control area) as shown in **Figure 9** and **Figure 10**. The apparent increase in lipid peroxidation may be attributed to the accumulation of the chemical pollutant in the organs under investigation indicating the existence of significant pollutant concentration in the fish liver and muscle of this area. Chemical pollutant catalyzes the formation of ROS capable to damage tissues such as DNA, proteins and lipids. [81]. The results of this study were in agreement with the work of Farombi *et al.*, [82] who found that the level of MDA activity in the liver of Cat fish (*Clarias gariepinus*) collected from Ogun River (polluted area) was significantly elevated in all fish organs compared to the control fishes collected from the Agodi fish farm, the percentage increase in liver, kidney, gills and heart lipid peroxidation were 177%, 102%, 168% and 71% respectively compared to control.

3.4. Microscopic Observations

3.4.1. Microscopic Observation of Hepatocytes

Ultrastructurally, it was noticed that hepatocytes of liver of fish "*S. rivulatus*" collected from El-Mex Bay, showed hepatic alternations. The present study pointed out nuclear alternations in "*Siganus rivulatus*" hepatocytes collected from a polluted area (El-Mex Bay) using transmission electron microscopy, including shrunk nuclei, centric nucleoli, decreased heterochromatin and irregular nuclear envelopes (**Figure 11**). Ultrathin preparations (electron microscope) of liver of "*S. rivulatus*" collected from El-Mex Bay showed also indistinct membrane of mitochondria with some of which exhibited partial loss of cristea, other exhibited swelling. One of the main alterations in liver cells of "*S. rivulatus*" collected from El-Mex Bay in the present study was the increase in size and number of lipid droplets (**Figure 11**). An increase, on the other hand, in the number of lysosomes, peroxisome proliferations and extensive vacuolization were observed in "*Siganus rivulatus*" from El-Mex Bay. Moreover, accumulation of vacuoles resulted in the displacement of nuclei to the cell margin with pyknosis of the nuclei and increased number of melanomacrophage centers in hepatocytes of "*S. rivulatus*" from El-Mex Bay was observed (**Figure 12**). Most of such ultra-structural changes included fragmentation of rough endoplasmic reticulum. Irregular nuclear outline and induction of glycogenosomes and phospholipids are of nonspecific nature.

3.4.2. Microscopic Observation of Muscle Fiber

Electron microscopic examination of the muscle fibers of "*S. rivulatus*" collected from the control area showed their composition of sarcoplasm and longitudinal arrays of myofibrils. The center of each dark (A-band or actin) was occupied by a pale area

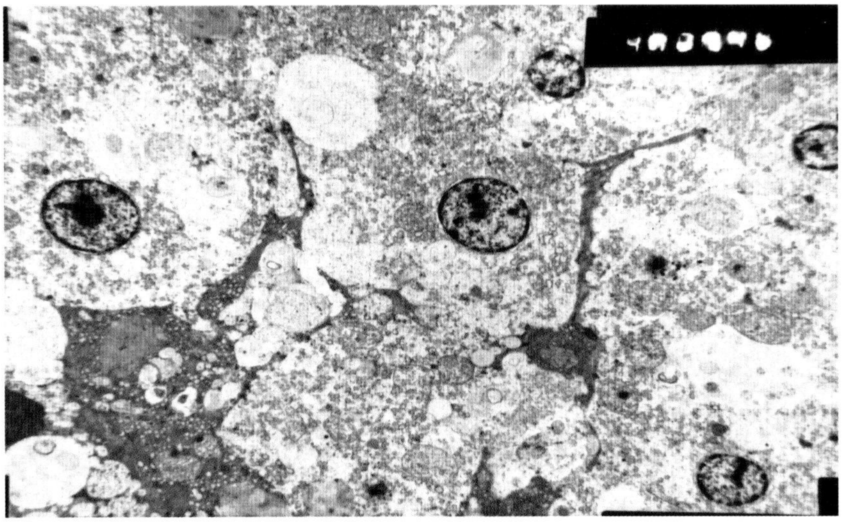

Figure 11. Electron micrograph. Liver section of "*Siganus rivulatus*" collected from Matrouh coast (control area). Showing, mononucleated hepatocytes interrupted by electron dense sinusoidal space (S), spherical central nucleus (N), with regular nuclear envelope (Ne), some nuclei showed centric large nucleoli (thin arrow), some nuclei revealed eccentric nucleoli (thick arrow), the hepatocyte nucleus possessed mostly regular nuclear envelope, small amount of condensed heterochromatin attached with the nucleolus (Nu), rough endoplasmic reticulum (RER), spherical and avoid mitochondria (M) with short tubular cristae, small number of lipid droplets (L) [Glutaraldehyde fixed-OsO_4 post fixed-uranyl acetate-lead citrate stained preparation ×6000].

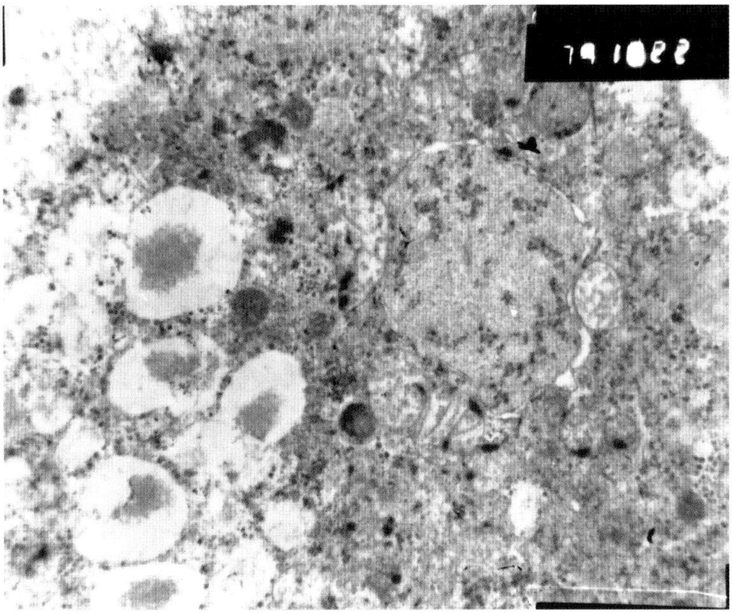

Figure 12. Electron micrograph, liver section of "*Siganus rivulatus*" collected from El-Mex Bay (tested area). Showing, altered nucleus (N) with irregular nuclear envelope (Ne), most chromatin aggregate peripherally, large number of microbodies (Mi), numerous primary lysosomes (Ly), perixosomes (P), large empty vacuoles (V) with flocculent materials and myline [Glutaraldehyde fixed-OsO_4 post fixed-uranyl acetate-lead citrate stained preparation ×10,500].

known as H-zone (Hensen's zone) which is bisected by an additional thin striation known as M-line (Mittelsheibe). A thin dark line, known as a Z-line (Zwischenscheibe) bisected the I-band which was relatively thick and distinguished by its density (**Figure 13**). Measurements based on electron microscopic preparations showed that there were insignificant differences in sacromere length, A-band, I-band and H zone in muscle fibres of "*Siganus rivulatus*" collected from control and test area at $p < 0.05$. Electron microscopic examination of the muscle of "*Siganus rivulatus*" collected from control area showed that the sarcoplasm is differentiated into peripheral sarcoplasm occupied a thin region and the nearest myofibrils, interfibrillar sarcoplasm filled the spaces between myofibrils and perinuclear sarcoplasm which is found at the pole of the nucleus. Moreover, the sarcoplasm contains non-myofibillar components including nuclei, mitochondria and sarcoplasmic reticulum, and cellular inclusions such as glycogen particles (**Figure 13**).

The nuclei of the "*Siganus rivulatus*" muscle collected from control area appeared just beneath the sarcolemma. They were oval or spindle shaped in appearance with predominant heterochromatin. Adjacent to the inner nuclear membrane, the heterochromatin appeared as dense aggregates surrounded by a regular nuclear envelope. The envelope appeared perforated by nuclear pores (**Figure 13**).

Electron microscopic preparations of the "*Siganus rivulatus*" muscle collected from tested area showed also that muscle fibers possessed highly altered nuclei, with large size nucleoli and irregular as well as disrupted nuclear envelope, with dilated nuclear pores (**Figure 14**). The analysis of the present study data using student t-test revealed that there is a highly significant decrease in the nuclear length in muscle cells of fish

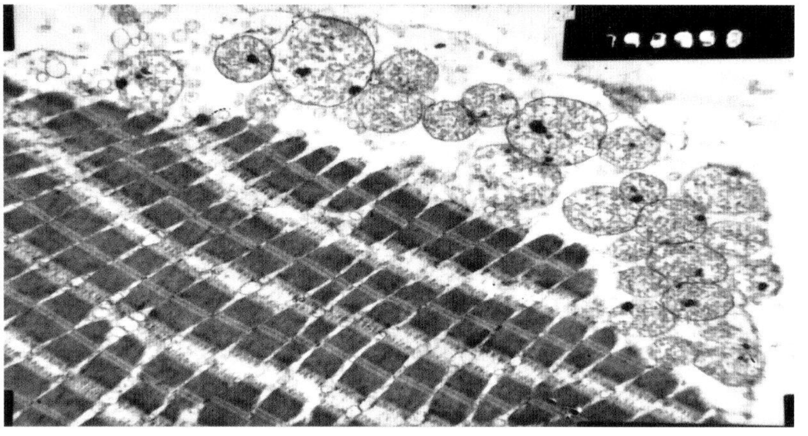

Figure 13. Electron micrograph. Longitudinal section of muscle of fish "*Siganus rivulatus*" collected from Matrouh coast (control area). Showing, regular striations, part of sarcolemma (Sl), numerous large size mitochondria (M) with distinct membrane, light matrix dense cristae. Arrow pointed at mitochondrial dense granules, sarcoplasmic reticulum (Sr). Note also, obvious isotropic band (I), Z-line (Z), anisotropic band (A), less distinct H-zone (H) [Glutaraldehyde fixed-OsO_4 post fixed-uranyl acetate-lead citrate stained preparation ×10,500].

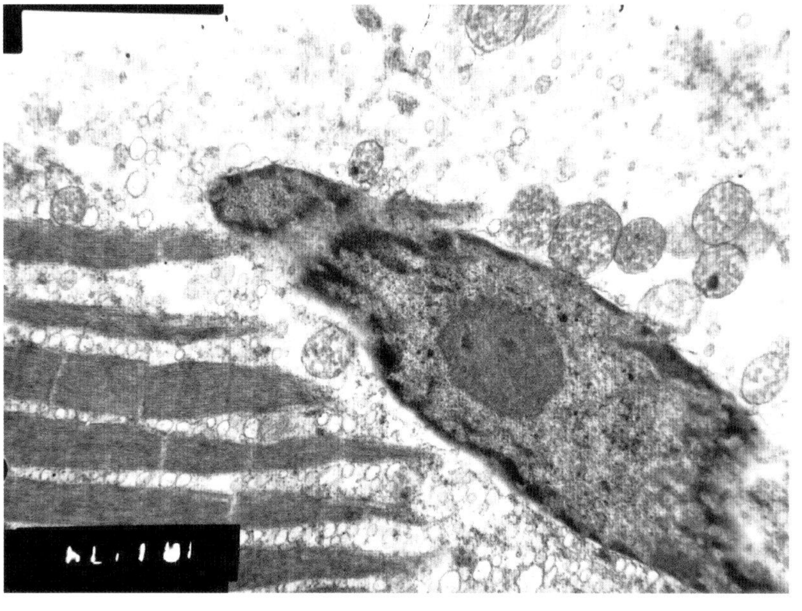

Figure 14. Electron micrograph. Longitudinal section of muscle of "*Siganus rivulatus*" collected from El-Mex Bay (tested area). Showing, highly altered nucleus (N), with altered large size nucleolus (Nu), with irregular disrupted nuclear envelope (Ne), with dilated nuclear pores (Np), small size mitochondria (M) with dense matrix, sarcoplasmic reticulum (Sr), increase lipid droplets (L), lysis in the cytoplasm [Glutaraldehyde fixed-OsO_4 postfixed-uranyl acetate-lead citrate stained preparation ×15,000].

caught from the test area comparing to its corresponding length in control area. However, there was insignificant difference in the nuclei width between the two area at $p < 0.05$.

Ultrastructural examination showed that the mitochondria of "*Siganus rivulatus*" muscle collected from the control area were oval in shape. They were clearly visible in the sarcoplasm adjacent to the nuclei and were also seen between the myofibrils. On the other hand, it was noticed that in "*Siganus rivulatus*" muscle collected from the test area, mitochondria appeared small in size with dense matrix. In these preparations the sub-sacrolemma mitochondria appeared disrupted and lysis in the cytoplasm (**Figure 14**). Data analysis using student t-test demonstrated that there is a highly significant decrease in the mitochondria length and width in the test area, comparing to that of control area at $p < 0.05$. In this context, the histopathological study confirmed the biochemical result where, the hepatocytes of liver of fish "*S. rivulatus*" collected from El-Mex Bay, showed hepatic alterations. It is suggested that this is due to increase concentration of different pollutants in El-Mex Bay [83]. This study also pointed out nuclear alterations in *S. rivulatus* hepatocytes collected from a polluted area using transmission electron microscope, including shrunk nuclei, centric nucleoli, decreased heterochromatin and irregular nuclear envelopes. It showed also mitochondrial indistinct membrane, some of which exhibited partial loss of cristea, other exhibited swelling. The mitochondrial degeneration may account for the impaired oxidative capability of hepatocytes in fish collected from the polluted location. According to the obtained biochemical results, the chemical pollutants induce lipid peroxidation and decrease the functional lipid content of tissues which can easily disturb the cellular metabolism and affect the ultra-structure. Thus, mitochondrial membranes damage affects their features; permeability, and selectivity. In consequence, they cannot regulate physiological processes and damaged mitochondria do not produce enough ATP, resulting in impair enzymatic detoxification within the exposed tissues [84]. These results are in agreement with results reported by Abdel-Moneim and Abdel-Mohsen in 2010, who found that catfish (Clariidae) hepatocytes collected from the polluted area showed accumulation of heterochromatin, enlarged nucleoli, and an extremely folded nuclear envelope. Tripathi and Shukla [85] Showed the mitochondrial degeneration may account for the impaired oxidative capability of hepatocytes in fish collected from the polluted area. Indeed, marked ultrastructural changes including, the presence of swollen mitochondria with loss of functional cristae due to exposed to methyl parathion. Electron microscopic examination of the *S. rivulatus* collected from test area showed also that muscle fibers (**Figure 14**) possessed highly altered nuclei with large size nucleoli, with irregular and disrupted nuclear envelop and mitochondria appeared small in size with dense matrix revealing that the histopathological biomarker of toxicity in fish organs is a useful indicator of environmental pollution.

4. Conclusions

The present study showed that El-Mex Bay area was polluted by different types of chemicals including hydrocarbons comparing with the reference area (Matrouh area). It was proven that the use of selected biochemical markers and chemical analysis has been a suitable to monitor the level of contamination of aquatic environment. The present study suggests that oxidative stress biomarkers, especially estimation of antioxidant systems in fish could provide a useful indicator of pollution of marine ecosystem. The induction of antioxidant systems in fish collected from polluted area in response to oxidative stress should be considered as a clear indication of the presence of pollution and environmental health degradation. The increase in lipid peroxidation revealed as a useful indicator of the pollution load in the present study. These results suggest also that chemical pollution is capable of inducing morphological alteration in liver of fish collected from polluted area. The present study indicated that ultrastructural changes serve as biomarker of stress in aquatic environment. When evaluating aquatic environmental pollution, it should always take into account a combination of several biochemical markers in any final assessment together of course with the results of chemical monitoring because only a combination of all this information will give the most objective picture of the status of the environmental monitoring.

References

[1] Wahby, S.D. and Shriadach, M.A. (1984) The Effect of Sewage Discharge on Some Chemical Characteristics of Seawater. *7th Journees Etud Pollutions*, Lvcerene, 11-13 October 1984, 81-90.

[2] Emara, H.I., Shriadah, M.A., Maoustafa, T.H. and El-Deek, M.S. (1995) Trace Metals-Nutrient Salts Relationship in Coastal Seawater of Alexandria. *Proceedings of the 2nd International Conference on the Mediterranean Coastal Environment*, Tarragona, October 24-27 1995, 1457-1464.

[3] Shriadah, M.A. and Emara, H.I. (1991) The Distribution of Chromium, Copper, Cadmium, and Lead in Areas of Multi-Polluting Factors of Alexandria. *International Proceeding of the Sympposium of Marine Chemistry in the Arab Region*, Suez, April 1991, 30-50.

[4] Shriadah, M.A., Shobier, A.H. and Shreadah, M.A. (2013) Assessment of Arsenic and Vanadium Pollution in Surface Sediments of the Egyptian Mediterranean Coast. *Journal of Environmental Technology and Management*, **16**, 82-101.

[5] Shriadah, M.A., Masoud, M., Khattab, A.M. and El Zokm, G. (2014) Impacts of Different Drains on the Seawater Quality of El-Mex Bay (Alexandria, Egypt). *Journal of Ecology and the Natural Environment*, **8**, 287-303.

[6] Shriadah, M.A., Abdel Fattah, L. and Fahmy, M.A. (2015) Heavy Metals in Some Fish Species and Bivalves from the Mediterranean Coast of Egypt. *Journal of Environmental Protection*, **6**, 1-9. http://dx.doi.org/10.4236/jep.2015.61001

[7] Shriadah, M.A., Shobier, A.H., Abdel Ghani, S., El-Zokm, G. and Said, T.O. (2015) Major Ions Anomalies and Contamination Status by Trace Metals in Sediments from Two Hot Spots along the Mediterranean Coast of Egypt. *Environmental Monitoring and Assessment*, **187**, 280. http://dx.doi.org/10.1007/s10661-015-4420-y

[8] Shriadah, M.A., Masoud, M., Khattab, A.M. and El Zokm, G. (2015) Nutrient Fluxes and Sediments Composition in El Mex Bay and Surround Drains, Alexandria, Egypt. *American Journal of Analytical Chemistry*, **6**, 513-527. http://dx.doi.org/10.4236/ajac.2015.66050

[9] Shriadah, M.A., Said, T.O., El-Sharkawi, F.M. and Mohamed, M.A. (2016) POPs in Sedi-

ments of the Eastern and Western Coast of the Egyptian Mediterranean Sea: A Comparative Study. *Development in Analytical Chemistry*, **3**, 1-11. http://dx.doi.org/10.14355/dac.2016.03.001

[10] Shriadah, M.A., Mohmoud, T.H., El-Deek, M.S. and Emara, H.I. (1990) Study and Interpretation of the Recent State of Pollution in the Eastern Harbor of Alexandria. *1st International Environmental Science Conference*, Sabha, 27-31 January 1990.

[11] Emara H.I. and Shriadah, M.A. (1991) Manganese, Iron, Cobalt, Nickel, and Zinc in the Eastern Harbor and El-Mex Bey Waters (Alexandria). *International Proceedings of the Symposium of Marine Chemistry in the Arab Region*, Suez, April 1991, 97-112.

[12] Shriadah, M.A. (1992) Trace Elements Concentration in Fish Samples from Alexandria Region. *Bulletin High Institute of Public Health*, **22**, 437-444.

[13] Shriadah, M.A. and Emara, H.I. (1996) Heavy Metals (Iron, Manganese, Nickel, Cadmium, and Lead) in the Sediments from the Eastern Harbor and El-Mex Bay of Alexandria, Egypt. *Proceedings of 6th International Symposium: Environmental Protection Is a Must*, Alexandria, 21-23 May 1996, 916-927.

[14] Abdel Fatah, L.M., Fahmy, M.A. and Shriadah, M.A. (2003) Zn, Cu, Cd, Pb and Hg in the Egyptian Coastal Sediments along the Mediterranean Sea. *Association for Modeling and Simulation in Enterprises*, **64**, 55-69.

[15] Shakweer, L., Shriadah M.A., Fahmy, M.A. and Abdel Fattah, L. (2006) Distributions and Concentrations of Trace Elements along the Mediterranean Coastal Water of Egypt. *Egyptian Journal of Aquatic Research*, **32**, 95-127.

[16] Abdel Ghani, S.A., Shobier, A.H., Said, T.O. and Shreadah, M.A. (2010) Organotin Compounds in Egyptian Mediterranean Sediments. *Egyptian Journal of Aquatic Research*, **36**, 221-229.

[17] Shreadah, M.A., Said, T.O., Abdel Moniem, M.I., Fathallah, E.M.I. and Mahmoud, M.E. (2011) PAHsin Sediments along the Semi-Closed Areas of Alexandria, Egypt. *Journal of Environmental Protection*, **2**, 700-709. http://dx.doi.org/10.4236/jep.2011.26081

[18] Shobier, A.H., Abdel Ghani, S.A. and Shreadah, M.A. (2011) Distribution of Total Mercury in Sediments of Four Semi-Enclosed Basins along the Mediterranean Coast of Alexandria. *Egyptian Journal of Aquatic Research*, **37**, 1-11.

[19] Leung, H.M., Leung, A.O.W., Wang, H.S., Maa, K., Liang, Y., Ho, K.C., Cheung, K., Tohidi, F. and Yung, K.K.L. (2014) Assessment of Heavy Metals/Metalloid (As, Pb, Cd, Ni, Zn, Cr, Cu, Mn) Concentrations in Edible Fish Species Tissue in the Pearl River Delta (PRD), China. *Marine Pollution Bulletin*, **78**, 235-245. http://dx.doi.org/10.1016/j.marpolbul.2013.10.028

[20] Shriadah, M.A. (1999) Oil Contamination along Oil Tanker Routes off the United Arab Emirates (The Arabian Gulf and The Gulf of Oman). *Bulletin of Environmental Contamination and Toxicology*, **63**, 203-210. http://dx.doi.org/10.1007/s001289900967

[21] Shriadah, M.A. (1999) Petroleum Hydrocarbons along the Arabian Gulf Coast of the United Arab Emirates. *Indian Journal of Marine Science*, **28**, 10-16.

[22] Shriadah, M.A. (2000) Levels and Distributions of Petroleum Hydrocarbons in the Coastal Waters and Sediments of the United Arab Emirates in the Arabian Gulf and the Gulf of Oman. *Water, Air and Soil Pollution*, **119**, 247-256. http://dx.doi.org/10.1023/A:1005192523876

[23] Shriadah, M.A. (2000) Contamination by Petroleum Hydrocarbons in Two Semi-Enclosed Areas along the Arabian Gulf Coast of the United Arab Emirates. *OEbalia*, **26**, 49-62.

[24] Shriadah, M.A. (2001) Petroleum Hydrocarbons Concentrations in Arabian Gulf Fish Tis-

sues. *Bulletin Environmental Contamination and Toxicology*, **67**, 560-567. http://dx.doi.org/10.1007/s001280160

[25] El Deeb, M.K., Said, T.O., El Naggar, M.H. and Shriadah, M.A. (2007) An Assessment of Pollution Status of Abu Qir Bay with Polycycic Aromatic Hydrocarbons (PAHs) and Their Possible Origins. *Proceedings of the 8th International Conference on the Mediterranean Coastal Environment*, Alexandria, 13-17 November 2007, 895-906.

[26] Shriadah, M.A., Said, T.O., Othman, I.M., Fathallah, E.M.I. and Mahmoud, M.E. (2012) Polychlorinated Biphenyls and Chlorinated Pesticides in Sediments along the Semi-Closed Areas of Alexandria, Egypt. *Journal of Environmental Protection*, **3**, 141-149. http://dx.doi.org/10.4236/jep.2012.32018

[27] Shriadah, M.A., Said, T.O., Othman, I.M., Fathallah, E.M.I. and Mahmoud, M.E. (2014) Organochlorinesin Seawater from Egyptian Mediterranean Coast of Alexandria. *Development in Analytical Chemistry*, **1**, 19-24.

[28] Alvarez, M.C., Franco, A., Perez-Dominguez, R. and Elliott, M. (2013) Sensitivity Analysis to Explore Responsiveness and Dynamic Range of Multi-Metric Fish-Based Indices for Assessing the Ecological Status of Estuaries and Lagoons. *Hydrobiologia*, **704**, 347-362. http://dx.doi.org/10.1007/s10750-012-1314-7

[29] Duke, O.F. (2008) Source Determination of Polynuclear Aromatic Hydrocarbons in Water and Sediment of a Creek in the Niger Delta Region. *African Journal of Biotechnology*, **7**, 282-285.

[30] Cheng, Z., Man, Y.B., Nie, X.P. and Wong, M.H. (2013) Trophic Relationships and Health risk Assessments of Trace Metals in the Aquaculture Pond Ecosystem of Pearl River Delta, China. *Chemosphere*, **90**, 2142-2148. http://dx.doi.org/10.1016/j.chemosphere.2012.11.017

[31] Sarkar, A., Ray, D., Shrivastava, A. and Sarker, S. (2006) Molecular Biomarkers: Their Significance and Application in Marine Pollution Monitoring. *Ecotoxicology*, **15**, 333-340. http://dx.doi.org/10.1007/s10646-006-0069-1

[32] Mohamed, B., Bouraoui, Z., Ghedira, J., Clearandeau, C., Jebali, J. and Boussetta, H. (2009) Seasonal Variation of Oxidative Stress Biomarkers in Clams *Ruditapes decussates* Sampled from Tunisian Coastal Areas. *Environmental Monitoring Assessment*, **155**, 119-128. http://dx.doi.org/10.1007/s10661-008-0422-3

[33] Luchmann, K., Toledo-Silva, G., Bainy, A. and Marques, M. (2007) Glutathione *S*-Transferase Cytosolic Isoform in the Pink-Shrimp, *Farfantepenaeus brasiliensis*, from Conceicao Lagoon, Santa Catarina Island, SC, Brazil. *Environmental International*, **33**, 546-549. http://dx.doi.org/10.1016/j.envint.2006.11.005

[34] Halliwell, B. and Gutteridge, J. (1999) Free Radicals in Biology and Medicine. Oxford University Press, Oxford.

[35] Stephensen, E., Sturve, J. and Forlin, L. (2002) Effects of Redox Cycling Compounds on Glutathione Content and Activity of Glutathione-Related Enzymes in Rainbow Trout Liver. *Comparative Biochemistry and Physiology Part C*, **133**, 435-442. http://dx.doi.org/10.1016/s1532-0456(02)00129-1

[36] Dusmant, M., Karmabeer, J., Navnath, M. and Tanaji, G. (2010) Assessment of Oxidative Stress Indices in a Marine Macro Brown Alga *Padina tetrastromatica* (Hauck) from Comparable Polluted Coastal Regions of the Arabian Sea. *Journal of Environmental Sciences*, **22**, 1413-1418. http://dx.doi.org/10.1016/S1001-0742(09)60268-0

[37] Fahmy, M.A., Tayel, F.R. and Shriadah, M.A. (1995) Effect of Pollution on the Water Quality of Mex Bay and Dekhaila Harbor of Alexandria. *1st International Conference on Present and Future Technology of Navigation and Marine Science of Mediterranean and the Red*

Sea, Alexandria, 29-31 October 1995.

[38] Shriadah, M.A. and Emara, H.I. (1992) Major Cations and Alkalinity in the Eastern Harbor and El-Mex Bey. *Bulletin Faculty of Science*, **32**, 156-174.

[39] Shriadah, M.A. and Emara, H.I. (1992) Iron, Manganese, Nickel, Lead, and Cadmium in Fish and Crustacea from the Eastern Harbor and El-Mex Bay of Alexandria. *Bulletin of High Institute of Public Health*, **22**, 515-525.

[40] Shriadah, M.A., Emara, H.I., Maoustafa, T.H. and El-Deek, M.S. (1992) Effects of Sewage and Industrial Wastes on the Chemical Characteristics of the Eastern harbor and El-Max Bay Waters of Alexandria, Egypt. *Science of the Total Environment*, **112**, 773-784.

[41] Said, M.A., El-Deek, M.S., Mamoud, T.H. and Shriadah, M.A. (1994) Effect of Pollution on the Hydrochemical Characteristics of Different Water Types in El-Mex Bay Area West of Alexandria, Egypt. *Acta Adriatica*, **34**, 9-19.

[42] Tayel, F.R., Fahmy, M.A. and Shriadah, M.A. (1996) Studies on the Physic Chemical Characteristics of Mex Bay and New Dekhaila Harbor Waters of Alexandria, Egypt. *Bulletin National Institute of Oceanography & Fisheries*, **22**, 1-18.

[43] Shriadah, M.A., Said, T.O., Younis, A.M. and Farag, R.S. (2006) Physico-Chemical Characteristics of the Semi-Closed Areas along the Mediterranean Coast of Alexandria, Egypt. *Egyptian Journal of Aquatic Research*, **32**, 38-48.

[44] Shriadah, M.A., Said, T.O., Younis, A.M. and Farag, R.S. (2006) Speciation of Organotin Compounds in Sediments of Semi-Closed Areas along the Mediterranean Coast of Alexandria. *Chemistry and Ecology*, **22**, 395-404. http://dx.doi.org/10.1080/02757540600917443

[45] Fathy, S.A., Abdel Hamid, F.F., Shreadah, M.A., Mohamed, L.A. and El-Gazar, M.G. (2012) Effect of Some Environmental Pollutants on Enzymatic and Total Antioxidant Activities in Tilapia Niloticus. *Blue Biotechnology Journal (BBJ)*, **1**, 433-443.

[46] Fathy, S.A., Abdel Hamid, F.F., Shreadah, M.A., Mohamed, L.A. and El-Gazar, M.G. (2012) Application of Principal Component Analysis for Developing Water Quality Index for Selected Coastal Areas of Alexandria Egypt. *Recourses and Environment Journal*, **2**, 297-305.

[47] Sukumaran, P. and Das, A. (2004) Distribution and Abundance of Rotifers in Relation to Water Quality of Some Tropical Reservoirs. *Indian Journal of Fisheries*, **51**, 295-301.

[48] Jangaard, P., Brockerhoff, H., Burgher, R. and Hoyle, R. (1967) Seasonal Changes in General Condition and Lipid Content of Cod from Inshore Waters. *Journal of the Fisheries Research Board of Canada*, **24**, 607-612. http://dx.doi.org/10.1139/f67-052

[49] UNEP/IOC/IAEA (1991) Sampling of Selected Marine Organisms and Sample Preparation for the Analysis of Chlorinated Hydrocarbons. Reference Methods of Marine Pollution Studies No. 12, Revision 2, United Nations Environment Programme, Nairobi, 17.

[50] Habig, W.H., Pabst, M.J. and Jokoby, W.B. (1974) Glutathione 5-Transferases. The First Enzymatic Step in Mercapturic Acid Formation. *Journal of Biological Chemistry*, **249**, 7130-7139.

[51] Beutler, E., Duron, O. and Kelly, B.M. (1963) Improved Method for the Determination of Blood Glutathione. *Journal of Laboratory and Clinical Medicine*, **61**, 882-888.

[52] Buege, J. and Aust, S. (1978) Microsomal Lipid Peroxidation. *Methods Enzymology*, **52**, 302-310. http://dx.doi.org/10.1016/S0076-6879(78)52032-6

[53] Reynolds, E. (1963) The Use of Lead Citrate at High pH as an Electron-Opaque Stain in Electron Microscopy. *Journal of Cell Biology*, **17**, 208-212. http://dx.doi.org/10.1083/jcb.17.1.208

[54] Kim, G.B., Maruya, K.A., Lee, R.F., Lee, J.H., Koh, C.H. and Tanabe, S. (1999) Distribution

and Sources of Polycyclic Aromatic Hydrocarbons in Sediments from Kyeonggi Bay, Korea. *Marine Pollution Bulletin*, **38**, 7-15. http://dx.doi.org/10.1016/S0025-326X(99)80006-X

[55] Witt, G. (1995) Polycyclic Aromatic Hydrocarbons in Water and Sediment of the Baltic Sea. *Marine Pollution Bulletin*, **31**, 237-248. http://dx.doi.org/10.1016/0025-326X(95)00174-L

[56] Yang, G.P. (2000) Polycyclic Aromatic Hydrocarbons in the Sediments of the South China Sea. *Environmental Pollution*, **108**, 163-171. http://dx.doi.org/10.1016/S0269-7491(99)00245-6

[57] Simpson, C.D., Mosi, A.A., Cullen, W.R. and Reimer, K.J. (1998) Composition and Distribution of Polycyclic Aromatic Hydrocarbon Contamination in Surficial Marine Sediment from Kitimat Harbor Canada. *Science of the Total Environment*, **181**, 265-278. http://dx.doi.org/10.1016/0048-9697(95)05026-4

[58] Salgueiro, L., Carballo, E., Falcón, M. and Gándara, J. (2009) Survey of Polycyclic Aromatic Hydrocarbons in Canned Bivalves and Investigation of Their Potential Sources. *Food Research International*, **42**, 983-988. http://dx.doi.org/10.1016/j.foodres.2009.04.003

[59] Sicre, M.A., Marty, J.C., Saliot, A., Aparicio, A., Grimalt, J. and Al-Baiges, J. (1987) Aliphatic and Aromatic in Different Sized Aerosols over the Mediterranean Sea: Occurrence and Origin. *Atmospheric Environment*, **21**, 2247-2259. http://dx.doi.org/10.1016/0004-6981(87)90356-8

[60] Soclo, H.H., Garrigues, P.H. and Ewald, M. (2000) Origin of Polycycilic Aromatic Hydrocarbons (PAHs) in Coastal Marine Sediments: Case Studies in Cotonou (Benin) and Aquitaine (France) Areas. *Marine Pollution Bulletin*, **40**, 387-396. http://dx.doi.org/10.1016/S0025-326X(99)00200-3

[61] El Sikaily, A., Khaled, A., El Nemr, A., Said, T.O. and Abd-Allah, A.M.A. (2002) Determination of Hydrocarbons in Bivalves from the Egyptian Mediterranean Coast. *Medcinal Marine Science*, **312**, 123-131.

[62] Nakata, H., Sakai, Y., Miyawaki, T. and Takemura, A. (2003) Bioaccumulation and Toxic Potencies of Polychlorinated Biphenyls and Polycyclic Aromatic Hydrocarbons in Tidal Flat and Coastal Ecosystems of the Ariake Sea, Japan. *Environmental Science & Technology*, **37**, 3513-3521. http://dx.doi.org/10.1021/es021083h

[63] Troisi, G., Bexton, S. and Robinson, I. (2006) Polyaromatic Hydrocarbon and PAH Metabolite Burdens in Oiled Common Guillemots (Uriaaalgae) Stranded on the East Coast of England (2001-2002). *Environmental Science & Technology*, **40**, 7938-7943. http://dx.doi.org/10.1021/es0601787

[64] Hylland, K., Tollefsen, K.E., Ruus, A., Jonsson, G., Sundt, R.C., Sanni, S., Utvik, T.I., Jangaard, P., Brockerhoff, H., Burgher, R. and Hoyle, R. (1967) Seasonal Changes in General Condition and Lipid Content of Cod from Inshore Waters. *Journal of the Fisheries Research Board of Canada*, **24**, 607-612. http://dx.doi.org/10.1139/f67-052

[65] Ahmad, I., Pacheco, M. and Santos, M. (2006) Oxidative Stress Biomarkers: An *in Situ* Study of Freshwater Wetland Ecosystem (Pateira de Fermentelos, Portugal). *Chemosphere*, **65**, 952-962. http://dx.doi.org/10.1016/j.chemosphere.2006.03.042

[66] Akcha, F., Izuel, C., Venier, P., Budzinski, H., Burgeot, T. and Narbonne, J.F. (2000) Enzymatic Biomarker Measurement and Study of DNA Adduct Formation in Benzo[a] Pyrene Contaminated Mussels, *Mytilus galloprovincialis*. *Aquatic Toxicology*, **49**, 269-287. http://dx.doi.org/10.1016/S0166-445X(99)00082-X

[67] Regoli, F., Gorbi, S., Frenzilli, G., Nigro, M., CorsiI, I., Focardi, S. and Winston, G.W. (2002) Oxidative Stress in Ecotoxicology: From the Analysis of Individual Antioxidants to a More Integrated Approach. *Marine Environmental Research*, **54**, 419-423.

http://dx.doi.org/10.1016/S0141-1136(02)00146-0

[68] Solé, M., Buet, A., Ortiz, L., Maynou, F., Bayona, J.M. and Albaigés, J. (2007) Biochemical Responses in Mussels Exposed to the Water-Accommodated Fraction of the "Prestige" Fuel Oil. *Scientia Marina*, **71**, 373-382. http://dx.doi.org/10.3989/scimar.2007.71n2373

[69] Güll, S., Belge-Kurutas, E., Yildiz, E., Sahan, A. and Doran, F., (2004). "Pollution Correlated Modifications of Liver Antioxidant System and Histopathology of Fish (*Cyprinidae*) Living in Seyhan Dam Lake, Turkey. *Environmental International*, **30**, 605-609. http://dx.doi.org/10.1016/S0160-4120(03)00059-X

[70] Zhen, Y. and Aili, J. (2011) Evaluation of Oxidative Stress Responses and Neurotoxicity Potential of Methamidophos in *Mytilus edulis*. *Advanced Materials Research*, **343-344**, 795-801. http://dx.doi.org/10.4028/www.scientific.net/AMR.343-344.795

[71] Kurutas, E.B., Sahan, A. and Altun, T. (2009) Oxidative Stress Biomarkers in Liver and Gill Tissues of Spotted Barb (*Capoeta barroisi* Lortet, 1894) Living in Theriver Ceyhan, Adana, Turkey. *Turkish Journal of Biology*, **33**, 275-282.

[72] Lesser, M.P. (2010) Survivorship, Oxidative Stress, and DNA Damage of Sea Urchin (*Strongylocentrotus droebachiensis*) Embryos and Larvae Exposed to Ultraviole Tradiation (290-400 nm) in the Gulf of Maine. *Photochemistry and Photobiology*, **86**, 382-388. http://dx.doi.org/10.1111/j.1751-1097.2009.00671.x

[73] Shahbudin, S., Deny, S., Zakirun, A.M., Haziyamin, T.A., Akbar John, B. and Taher, M. (2011) Antioxidant Properties of Soft Coral *Dendronephthya* sp. *International Journal of Pharmacology*, **7**, 263-267. http://dx.doi.org/10.3923/ijp.2011.263.267

[74] Ahmad, I., Pacheco, M. and Santos, M.A. (2004) Enzymatic and Non-Enzymatic Antioxidants as an Adaptation to Phagocyte-Induced Damage in *Anguilla Anguilla* L. Following *in Situ* Harbor Water Exposure. *Ecotoxicology and Environmental Safety*, **57**, 290-302. http://dx.doi.org/10.1016/S0147-6513(03)00080-0

[75] Santos, M.A., Pacheco, M. and Ahmad, I. (2004) *Anguilla Anguilla* L. Antioxidants Responses to *in Situ* Bleached Kraft Pulp Mill Effluent Outlet Exposure. *Environmental International*, **30**, 301-308. http://dx.doi.org/10.1016/S0160-4120(03)00178-8

[76] Ahmad, I., Oliveira, M., Pacheco, M.M, and Santos, M.A. (2005) *Anguilla Anguilla* L. Oxidative Stress Biomarkers Responses to Copper Exposure with or without β-naphthoflavone Pre-Exposure. *Chemosphere*, **61**, 267-275. http://dx.doi.org/10.1016/j.chemosphere.2005.01.069

[77] Havelkova, M., Blahova, J., Kroupova, H., Randak, T., Leontovycova, I., Grabic, R., Pospisil, R. and Svobodova, Z. (2008) Biomarkers of Contaminant Exposure in Chub (*Leuciscuscephalus* L.)—Biomonitoring of Major Rivers in the Czech Republic. *Sensors*, **8**, 2589-2603. http://dx.doi.org/10.3390/s8042589

[78] Siroka, Z., Krijt, J., Randak, T., Svobodova, Z., Peskova, G., Fuksa, J., Hajslova, J., Jarkovsky, J. and Janska, M. (2005) Organic Pollutant Contamination of the River Elbe as Assessed by Biochemical Markers. *Acta Veterinaria Brno*, **74**, 293-303. http://dx.doi.org/10.2754/avb200574020293

[79] Rathod, V. and Balkrishna, P.A. (2011) Lipid Peroxide, Glutathione and Glutathione-Dependent Enzyme (GST) in Mixed Zooplankton from the North-West Coast of India: Implication for the Use of Environmental Monitoring. *African Journal of Biotechnology*, **10**, 7862-7867. http://dx.doi.org/10.5897/AJB10.1909

[80] Havelková, M., Randák, T., Blahová, J. and Slatinská, I. (2010) Biochemical Markers for the Assessment of Aquatic Environment Contamination. *Interdisciplinary Toxicology*, **1**, 169-181.

[81] Pandey, S., Parvez, S., Sayeed, I., Haque, R., Bin-Hafeez, B. and Raisuddin, S. (2003) Biomarkers of Oxidative Stress: A Comparative Study of River Yamuna Fish *Wallago attu* (Bl. & Schn.). *Science of the Total Environment*, **309**, 105-115. http://dx.doi.org/10.1016/S0048-9697(03)00006-8

[82] Farombi, E., Adelowo, O. and Ajimoko, Y. (2007) Biomarkers of Oxidative Stress and Heavy Metal Levels as Indicators of Environmental Pollution in African Cat Fish (*Clarias gariepinus)* from Nigeria Ogun River. *International Journal Environmental Research Public Health*, **4**, 158-165. http://dx.doi.org/10.3390/ijerph2007040011

[83] Mohamed, F.A. (2001) Impacts of Environmental Pollution in the Southern Region of Lake Manzalah, Egypt, on the Histological Structures of the Liver and Intestine of *Oreochromis niloticus* and *Tilapia zillii*. *Journal of Egyptian Academy Society for Environmental Development*, **2**, 25-42.

[84] Ptashynski, M., Pedlar, R., Evan, R., Baron, C. and Klaver, J. (2002) Toxicology of Dietary Nickel in Lake White Fish (*Coregonus clupeaformis*). *Aquatic Toxicology*, **58**, 229-247. http://dx.doi.org/10.1016/S0166-445X(01)00239-9

[85] Tripathi, C.L. and Shukla, S.P. (1990) Enzymatic and Ultrastructural Studies in a Freshwater Catfish: Impact of Methyl Parathion. *Biomedical Environmental Science*, **3**, 166-182.

2

Optimizing Hydraulic and Chemical Properties of Iron and Aluminum Byproducts for use in On-Farm Containment Structures for Phosphorus Removal

Sheri L. Herron[1], Andrew N. Sharpley[2*], Kristofor R. Brye[2], David M. Miller[2]

Ecosustain, LLC, Dripping Springs, TX, USA
Department of Crop, Soil, and Environmental Sciences, University of Arkansas, Fayetteville, AR, USA
Email: *sharpley@uark.edu

Abstract

Phosphorus (P) transported in runoff from broiler production areas is a potential source for nutrient enrichment of local surface waters. Capturing nutrients prior to runoff leaving the production area could reduce the risk of transport. Commercially available systems for nitrogen (N) and P removal from stormwater are available but too expensive for widespread adoption on small, privately owned farms. The P adsorption capacities and hydraulic conductivities of locally-sourced iron red mud (RM) and aluminum water treatment residual (WT) were determined for potential on-farm treatment use. The byproducts were air-dried and separated into ≤2-, ≤6-, and ≤12.5-mm particle size fractions. Saturated hydraulic conductivity (HC) was determined for each particle-size fraction and results evaluated for the material's ability to transmit a 25-year, 30-min rainfall of 6.6 cm. While the HC of ≤6-mm particles of each byproduct did not differ ($p > 0.05$; 4.1 and 4.6 cm·min^{-1}), for ≤12.5-mm particles it was greater ($p > 0.05$) for WT than RM (15.4 and 8.0 cm·min^{-1}, respectively). However, all byproduct size fractions transmitted flow sufficiently to accommodate the baseline storm. Phosphorous adsorption maxima of ≤2-mm fraction for RM and WT were 25 and 10 g·P·kg^{-1}, respectively. Using a solution containing 6 mg·P·L^{-1}, representing the measured runoff-P concentration from areas adjacent to poultry house ventilation fans, the ≤2-mm fraction of RM removed 98% of added P and WT 84%. The ≤6-mm fraction of RM and WT removed 56% and 57% of added P, respectively, while the ≤12.5-mm fraction adsorbed only 28% and 10%. Results indicate the potential use of low-cost RM and WT byproducts to sequester P in runoff prior to P leaving poultry farm production areas.

Keywords

Phosphorus Runoff, Poultry Production, Red Muds, Water Quality, Water Treatment Residual

1. Introduction

Water quality impairment from phosphorus (P) enrichment is a major concern in many areas of the U.S., with agriculture purported to be a major contributor [1] [2]. Runoff from the land directly around broiler production facilities has been shown to be enriched with nutrients that could potentially contribute to nearby surface water quality impairment [3] [4]. Removing P from runoff water prior to it leaving the production area is more cost-effective than treating the receiving water once impaired [5]. While P can be removed during movement along a grassed waterway between poultry houses, P can accumulate in the soil, which can then represent a long-term source of P for off-site transport in runoff [6]. Removal of dissolved P from surface flow is more challenging than for particulate P and most remedial measures for dissolved P focus on removal of P as close to its source as possible [7] [8]. Therefore, conservation practices that trap P prior to the P leaving the production area are needed for protection of surface water quality.

Several P removal systems, primarily designed for municipal and construction site stormwater treatment, are commercially available [9] [10]. However, their use on the over 1600 privately owned poultry houses in the Illinois River watershed in northwest Arkansas and northeast Oklahoma is cost-prohibitive [11]. The effectiveness of the commercial systems is limited by their low hydraulic conductivity, which can restrict flow through the P-removal media during high-flow events. The coarser particle-size fraction used in the overflow weir design is adequate for hydraulic transmission, but results in reduced reaction time, which can limit P removal from solution [12].

Materials specifically manufactured to adsorb P tend to be costly and are, therefore, impractical for use at a broiler production facility. However, industrial byproducts with high P-adsorption capacities do exist. Water treatment residuals (WT) are one such class of byproducts. They are produced in large quantities by municipal drinking water treatment plants when coagulants, such as alum, are added to raw water to flocculate solids suspended in the water. A local drinking water treatment plant in northwest Arkansas uses alum as a flocculent and the resulting "floc" is skimmed from the treated water and centrifuged to remove free water, then disposed of in landfills [13].

Another locally-sourced product of potential use is an iron filter cake generated in the manufacturing process of steel belts for steel-belted, radial-ply tires. During the steel-belt production process, wastewater is created which contains ferrous iron, hydrochloric acid, and phosphoric acid. The pH of the wastewater is adjusted to greater than 8.5 with calcium hydroxide to create metal hydroxides, after which an ionic polymer is added to flocculate the metal hydroxides. Water is removed from the flocculent to create a red mud (RM) filter cake, which is disposed of in landfills.

Byproducts similar to RM and WT have been shown to be potentially useful in enhancing P removal in constructed wetlands [12], as once dried, these byproducts have high P adsorption capacities [14] [15]. The objective of this study was to evaluate the P-adsorption and water transmission characteristics of WT and RM byproducts that may have potential use as P-adsorbing materials in on-farm P removal systems. It was hypothesized that the WT and RM byproducts would have large P-adsorption capacities. However, WT contains P due to its use as an adsorbent to remove P in the drinking-water filtration process, and RM contains P due to its use as a cleaner in the manufacturing process to produce steel belts. Additionally, it was hypothesized that larger particle-size fractions of each of the byproducts would have greater HC and lower P-sorbing capacities than smaller particle-size fractions.

2. Materials and Methods

2.1. Byproduct Selection and Particle Sizing

Red mud, iron filter cake was obtained from a local manufacturing company and water treatment (WT) residual was obtained from a water treatment plant in northwest Arkansas. The byproducts were loaded onto a tarp-lined flatbed trailer. The WT and RM were sieved to remove material >12.5 cm and were air-dried. As both P adsorption and hydraulic conductivity were expected to be affected by byproduct particle size, particle-size fractions were subsequently selected to contain both small particle sizes for maximum P adsorption and larger particle sizes to allow for sufficient hydraulic conductivity. Screen sizes with 6.35-mm (0.25 inch) and 12.7-mm (0.5 inch) openings were chosen based on availability of wire mesh to allow processing of large volumes of material. The byproducts shrink during the drying process, resulting in a mixture of particle sizes ranging up to the mesh size of the screen. Additional crushing was required to generate the ≤2-mm-sized material. Once dried, samples were sieved into particle-size fractions of ≤2-, ≤6-, and ≤12.5-mm. The particle-size distribution of RM and WT in the three size classes were determined by sequential dry sieving (**Table 1**).

Polycarbonate columns 25.4-cm long and 7.62 cm in diameter were constructed (**Figure 1**). Inlet and outlet holes were drilled 2 cm from the bottom and top of column, respectively. Rubber stoppers were glued to the bottom of each column. One end of a piece of 0.95-cm-diameter tubing was attached to the inlet connector and the other end to a funnel mounted on a stand at a height to create 5 cm of hydraulic head. Another section of 0.95-cm-diameter tubing was attached to the outlet connector and the end of the tubing directed into a collection container. An 18-L reservoir was used to supply flow, regulated with a hose clamp, to the funnel to maintain a constant head.

2.2. Chemical Analyses

Initial total N concentrations of RM and WT were determined by dry combustion (Nelson and Sommers, 1996). Initial total P (TP) and total K were determined by inductively coupled, argon-plasma, optical emission spectroscopy following nitric acid

Table 1. Particle-size distribution (by volume) of red mud (RM) iron filter cake and water treatment (WT) residual byproduct materials.

Byproduct material/Size fractions	≤2 mm		≤6 mm		≤12.5 mm	
	cm^3	% of total	cm^3	% of total	cm^3	% of total
Red mud iron filter cake						
8 - 12.5 mm	0.0	0	0.0	0	1592.0	32
6 - 8 mm	0.0	0	0.0	0	1530.5	31
4 - 6 mm	0.0	0	902.1	30	563.0	11
2 - 4 mm	0.0	0	831.4	27	563.0	11
<2 mm	1202.9	100	1291.3	43	688.2	14
Total	1202.9	100	3024.8	100	4936.8	100
Water treatment residual						
8 - 12.5 mm	0.0	0	0.0	0	1592.0	31
6 - 8 mm	0.0	0	0.0	0	1611.7	31
4 - 6 mm	0.0	0	1132.1	31	638.4	12
2 - 4 mm	0.0	0	902.1	25	572.3	11
<2 mm	2211.1	100	1645.1	45	715.4	14
Total	2211.1	100	3679.3	100	5129.8	100

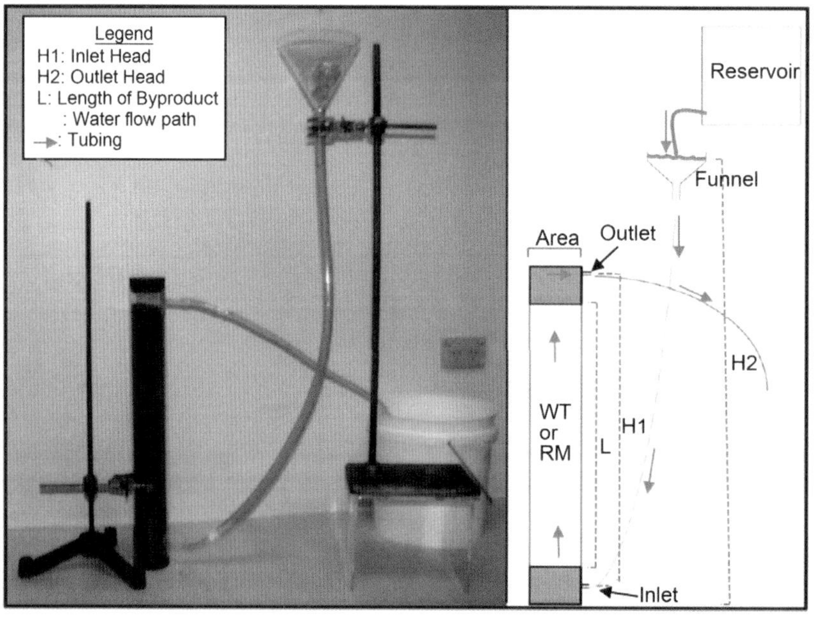

Figure 1. Column design for phosphorus adsorption column study, also used for hydraulic conductivity determinations (adapted from [16]).

digestion (PerkinElmer Optima 7300, Waltham, MA).

Phosphorus adsorption properties of ≤2 mm-sized RM and WT were determined by batch equilibrations, where 1 g of air dried byproduct was shaken end-over-end (60 rpm) with 40-mL distilled deionized water containing P of varying concentrations (0 to 1000 mg·L^{-1} as K_2PO_4) for 60 min at 25˚C. After shaking, samples were centrifuged for 10 minutes at 5250 x *g* and 25˚C, decanted, and filtered through a 0.45-μm membrane filter. Dissolved reactive phosphate (DRP) concentrations of filtrates were determined by the colorimetric molybdenum-blue method of Murphy and Riley [17]. Equilibrations were conducted in triplicate and adsorbed P calculated as the amount of P added minus the amount remaining in solution after shaking. Phosphorus adsorption isotherms were constructed by plotting the equilibrium solution P concentration (C) and P adsorbed per unit weight of byproduct (x/m). Phosphorus adsorption maxima (P_{max}) and binding energy (B) of RM and WT were calculated used the Langmuir isotherm equation [Equation (1)] [18]:

$$x/m = B * P_{max} * C/(1 + B * C) \quad (1)$$

2.3. Hydraulic Properties

Hydraulic conductivities (HC) of the byproducts were determined for each of the three particle-size fractions using a modification of the constant-head method to determine HC of saturated soils [16]. This technique was designed to estimate P release from RM and WT residuals (**Figure 1**) [19], but was modified to be used to also measure the HC of the two materials. The materials in each column were initially saturated with water until 5 mL were collected from the outlet of each column. Following initial saturation, a 5-cm constant head of pressure was used to input distilled water via the inlet at the bottom of the column. This method was chosen because, once dried, the RM and WT byproducts were hydrophobic and did not slake when wetted, making it difficult to remove air pockets and fully saturate the material when saturated from above, thus saturation from below was necessary.

Early attempts to saturate material from the bottom of the column and then apply a constant head from the top of the column were unsuccessful due to blockage by air pockets. Glass wool was then placed in the bottom of the column covering the inlet. A rubber ring and a wire screen were placed on top of the glass wool to hold the glass wool in place and create a level bottom, on top of which was placed 8 cm of byproduct. The alternative method allowed for improved flow and for a more consistent measurement of material HC, which closely represented field conditions. Red mud and WT were packed in columns to a constant bulk density (**Table 2**). Another layer of glass wool, sandwiched between two wire screens, was placed on top of the byproduct, below the outlet, to eliminate particle floating.

Hydraulic conductivity determinations were performed in triplicate on the three particle-size fractions of the RM and WT byproducts with water flow for 10 min. A constant head pressure was maintained using an elevated 18-L reservoir of water. Constant head hydraulic conductivity was calculated as:

$$HC = (V * L)/(A * t * H) \quad (2)$$

where HC is the saturated hydraulic conductivity (cm·min^{-1}), V is the volume of flow

Table 2. Bulk density of different particle-size fractions of iron (RM) and aluminum (WT) byproduct materials.

Byproduct material	Particle-size fraction	Bulk density
	mm	$g \cdot cm^{-3}$
	≤2	0.82
Red mud iron filter cake	≤6	0.80
	≤12.5	0.87
	≤2	0.85
Water treatment residual	≤6	0.86
	≤12.5	0.84

for a 10-min run (cm^3), L is length of byproduct in column (cm), A is the internal area of column (cm^2), t is the time of run (min), and H is the hydraulic head determined by H2-H1 from **Figure 1** (cm).

2.4. P Adsorption Properties

The same particle-size fractions (≤2, ≤6, and ≤12.5 mm) of RM and WT used in the HC determinations were used in a P adsorption column study. This experiment was designed to simulate field conditions, where a P-containing solution flows through the byproduct. Phosphorus adsorption in this system was expected to be lower than that observed in traditional batch equilibrations, which had a longer contact time between P in solution and P-sorbing WT or RM materials (*i.e.*, 60 and 10 min, respectively). The 18-L reservoir supplying the constant-head flow to the columns had a concentration of 6 $mg \cdot P \cdot L^{-1}$, which was based on the average flow-adjusted runoff TP concentration from simulated rainfall plots adjacent to poultry house fans. Treatments were conducted in triplicate on the three particle-size fractions of both RM and WT byproducts. Outflow solutions were subsampled for analysis at the beginning of flow and every two minutes thereafter for a total of 30 min. Samples were filtered through a 0.45-μm membrane filter and DRP concentration was determined by the colorimetric molybdenum-blue method of Murphy and Riley [17].

2.5. Statistical Analysis

A randomized block design with three replications for each P concentration tested was used to evaluate P adsorption of each byproduct at different solution-P concentrations. A one-factor analysis of variance (ANOVA) was conducted on results from the batch equilibration study for each byproduct. A homogeneity of variance test was conducted to compare HC of the ≤6- and ≤12.5-mm particle-size fractions of RM and WT. A homogeneity of variance test was also conducted to compare P removal of the ≤2-, ≤6-, and ≤12.5-mm particle-size fractions of RM and WT. Since variances were not homogenous for either HC or P removal, which violated the requirements for a two-factor ANOVA separate one-factor ANOVAs were conducted to evaluate the effect of byproduct on HC and effect of particle size on HC and to evaluate the effect of byproduct on P-removal capacity and effect of particle-size fraction on P-removal capacity. Significance was judged at the $p < 0.05$ level unless noted otherwise. Wolfram Mathematica (online beta version 2014, Wolfram, Champaign, IL) was used to conduct all statistical analyses.

3. Results and Discussion

3.1. Byproduct Composition

The initial TP concentration of the RM and WT was 33.9 and 1.2 mg·TP·g^{-1}, respectively (**Table 3**). However, initial P concentrations were not a concern due to the fact that P adsorbed by these residuals is not released unless the material is acidified (*i.e.*, dissolution of Fe and Al compounds) or becomes anaerobic (*i.e.*, reductive dissolution of Fe compounds) [12] [14]. Initial N and K concentrations in the RM were below detection limits of 1.0 mg·kg^{-1} for N and 100 mg·kg^{-1} for K. The WT initially contained 7700 mg·total·N·kg^{-1} and 1400 mg·K·kg^{-1}. Both byproducts had a pH of 7.0 (**Table 3**).

3.2. Isotherms

Phosphorus adsorption determined by batch equilibrations reflects the remaining capacity of RM and WT to adsorb further P. Adsorption capacities between replicates did not differ ($p > 0.05$; **Table 4**). Over 90% of added P was removed by RM and WT, up to P applications of 13 and 3 mg·P·g^{-1} of byproduct, respectively (**Table 4**). The linearized form of the Langmuir equation was used to calculate P adsorption maxima and binding energies (see Equation (1)) (**Figure 2**):

$$C/x/m = \left(1/B * P_{max}\right) + C/P_{max} \tag{3}$$

A plot of C/x/m against C should be a straight line if the Langmuir equation describes the isotherm, and the slope of the line is equal to $1/P_{max}$. As shown in **Figure 3**, the Langmuir equation described these isotherms ($R^2 > 0.96$). The calculated P_{max} was 25 and 10 mg·P·g^{-1} for RM and WT, respectively. **Table 5** summarizes previously reported P_{max} values for several P-sorbing materials, which range from 0.0003 mg·P·g^{-1} for limestone to 31.97 mg·P·g^{-1} for West Virginia ochre, an iron mining byproduct. The P_{max} value for RM was greater than 87% of the materials listed and that for WT was greater than 74% of those listed. Clearly, the locally sourced RM and WT byproducts retain a large capacity to adsorb further P from solution.

Table 3. Initial nutrient composition and pH of red mud (RM) iron filter cake and water treatment residual (WT) materials.

Byproduct	Composition			pH
	Nitrogen	Phosphorus	Potassium	
	mg·kg^{-1}	mg·g^{-1}	mg·kg^{-1}	
Red mud iron filter cake	<1.0	33.9	<100.0	7.0
Water treatment residual	7700	1.2	1400	7.0

Table 4. Phosphorus removed from solution during batch equilibration of the ≤2-mm particle size red mud (RM) and water treatment (WT) residual materials.

Byproduct material	P added	P added	Average P removed	Average P removed	p-value*
	mg·L^{-1}	----- mg -----		%	
Red mud iron filter cake					
	60	2.4	2.37	98.7	
	90	3.6	3.55	98.7	
	120	4.8	4.74	98.7	
	150	6.0	5.82	96.9	
	200	8.0	7.73	96.7	
	250	10.0	9.53	95.3	0.98
	325	13.0	12.07	92.8	
	425	17.0	15.08	88.7	
	500	20.0	17.15	85.8	
	1000	40.0	24.02	60.1	
Water treatment residual					
	50	2.0	1.8	92.1	
	75	3.0	2.8	92.4	
	100	4.0	3.5	86.6	
	150	6.0	4.5	74.9	
	200	8.0	6.0	75.2	0.97
	250	10.0	7.3	73.1	
	300	12.0	7.5	62.6	
	350	14.0	8.6	61.4	

*p-value result of comparison across replicates for each byproduct.

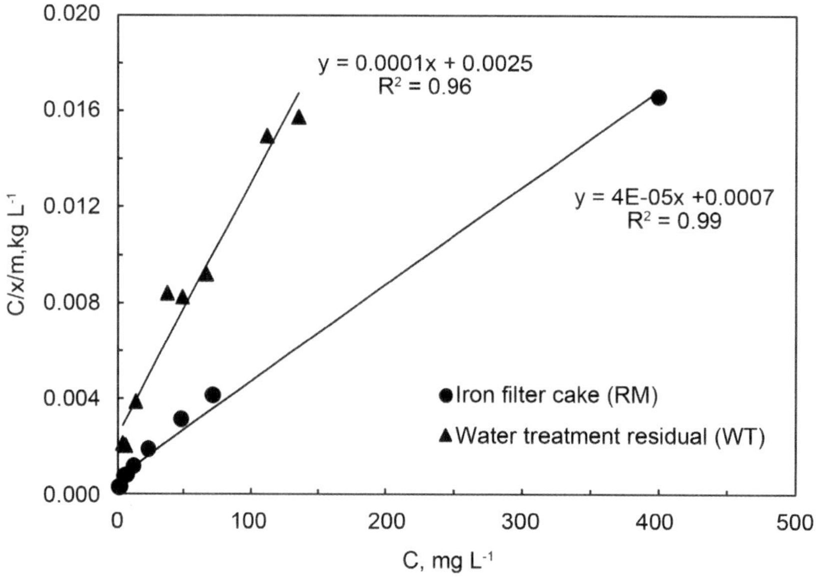

Figure 2. Description of the isotherms by the linearized Langmuir equation.

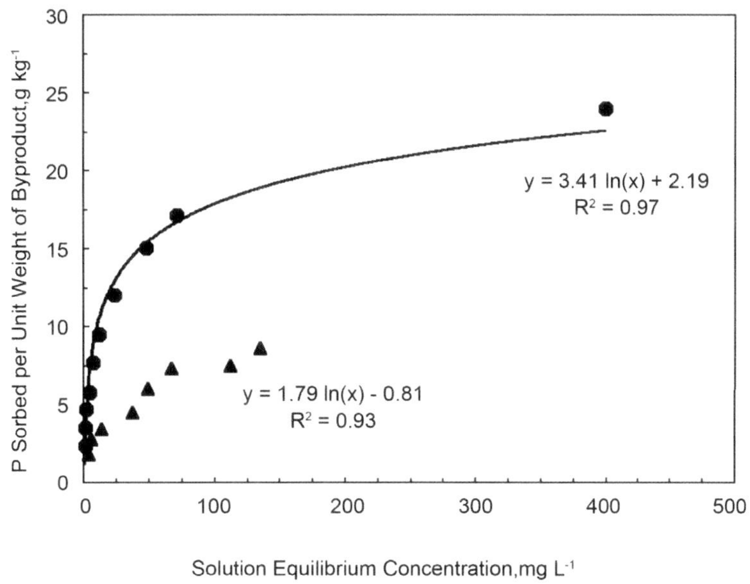

Figure 3. Phosphorus adsorption isotherms for ≤2-mm particle-sized byproducts.

3.3. Hydraulic Conductivity

Variability in HC for all byproduct particle-size fractions was observed between replicates as a result of particle settling in the columns. The HC was greater in 60% of the runs in the first 10-min replicate than the subsequent two 10-min replicates for all particle sizes of RM. The HC replications varied by 3.4, 3.1, and 10.5 $cm{\cdot}min^{-1}$ for RM particle-size fraction treatments of ≤2-, ≤6-, and ≤12.5-mm, respectively. The HC was greater in the first 10-min replicate than the last two 10-minute replicates in 100% of the runs for all particle sizes of WT. The HC replications varied by 0.6, 3.2, and 56.6 $cm{\cdot}min^{-1}$ for WT particle-size fraction treatments of ≤2-, ≤6-, and ≤12.5-mm, respectively. Due to the variability in HC for all byproduct treatments, the first 10-minute replications were treated as conditioning runs to allow for byproduct settling and were excluded from the final calculations of HC for RM and WT.

Hydraulic conductivity varied notably for both the RM and WT ≤12.5-mm particle-size fraction treatments, ranging from 8.3 to 25.7 $cm{\cdot}min^{-1}$ for RM and from 15.7 to 75.0 $cm{\cdot}min^{-1}$ for WT (**Table 6**). The greater variability in HC with larger byproduct particle sizes was partially attributed to variability in capillary flow due to finer particles settling and clogging flow pathways, which was observed when material was removed from the columns following completion of the study. The ≤2-mm particle-size fraction treatment for both byproducts showed visibly dry material in the interior of the column following the study, indicating that they did not fully saturate in the column, thus were excluded from further analysis. This was of limited concern as the ≤2 mm material is too fine to be of value in the field by itself as a P-adsorbing residual receiving large volumes of water.

Comparison of HCs for the ≤6- and ≤12.5-mm particle-size fraction treatments yielded no difference ($p > 0.05$) in HC between ≤6- and ≤12.5-mm RM (**Table 6**). However, there was a significant difference ($p = 0.02$) in HC for the WT, with the ≤12.5-mm

Table 5. Maximum phosphorus (P) adsorption capacities of a variety of materials that have been considered as P adsorbents.

Substrate	P-adsorption maximum $mg \cdot P \cdot g^{-1}$	Reference
Ace ALD ochre	7.41	[21]
AMD flocs – Lime treated	0.73	[22]
AMD flocs – NaOH treated	0.76	[22]
AMD flocs – NH_4 treated	0.86	[22]
Apatite	0.28 - 1.09	[23]
Avoca ochre	21	[24]
Babb Creek ochre	1.82	[21]
Bauxite	0.61	[25]
Blast Furnace Slag	0.40 - 0.45	[26]
Burnt oil shale	0.65	[25]
Coal ash	0.081 - 29.5	[27]
Dolomite	0.168	[28]
Dolomite sand	0.417	[29]
Fly Ash	0.62	[26]
Fly ash	5.5 - 42.6	[30]
Friendship Hill ochre	23.9	[21]
Gravel	0.03 - 0.05	[26]
Iron ochre (coal mine drainage treatment)	26	[31]
Iron-coated sand	27.4	[14]
Laterite	0.75	[32]
Minto Ochre	30.5	[33]
Ochre pellets	22	[31]
Polkemmet Ochre	26	[33]
Randy Camp ochre	5.53	[21]
Sands (Danish)	0.02 - 0.129	[34]
Shale	0.75	[32]
Shell sand	14 - 17	[35]
Shell sand	0.8 - 8.0	[36]
Shell sand	9.6	[37]
Steel Slag	0.38	[26]
Toby Creek ochre	22.6	[21]
West Virginia ochre	31.97	[38]
Zeolite	1	[32]
Zeolite	2.15	[39]

material five-fold greater than that for the ≤6-mm material (**Table 6**). Further, the HC of RM and WT was similar ($p > 0.05$) for both size fractions.

Table 6. Average and variance of hydraulic conductivity of red mud (RM) iron filter cake and water treatment (WT) residual as a function of byproduct particle-size fraction.

Byproduct material	Particle-size fraction	Hydraulic conductivity		*p*-value*
		Average	Variance	
	mm	cm·min^{-1}		
	Red mud iron filter cake			
	<6	10.6	14.6	0.14
	<12.5	16.3	45.9	0.02
	Water treatment residual			
	<6	8.0	4.1	0.44
	<12.5	43.1	684.8	0.05

**p*-value result of comparison across replicates for each byproduct.

The HC of the ≤6- ≤12.5-mm RM and WT materials all exceeded the capacity to transmit rainfall during a 25-yr, 30-min storm (6.6 cm, 0.22 cm·min^{-1}) [20]. For the ≤6 mm WT material, HC was greater than 5.5 cm·min^{-1} in all but one replicate. These results suggest that both RM and WT material of ≤6-mm and ≤12.5-mm size fractions could accommodate water flow from a 25-yr, 30-min rainfall event.

3.4. P Adsorption Column Study

Following HC measurement, the columns were used to estimate P adsorption, although applying a constant-head flow simulated extreme high flow conditions and resulted in reduced contact time between P solution and byproduct. The amount of solution that passed through the columns during the 30-min sampling period varied with particle size. For the ≤2-mm material, half the solution volume was applied compared to ≤6-mm material, allowing for additional contact time between solution P and byproduct particles. The larger reactive surface area of ≤2-mm-sized material compared to other material sizes resulted in 98% P removal by RM and 94% P removal for WT byproduct (**Table 7**).

While total flow volume for the ≤12.5-mm RM and WT material was greater than for ≤6-mm material, there was significantly less surface area and less added-P sorbed by RM (30% of added P) and WT (11% of added P) (**Table 7**). Phosphorus removal by the ≤2-, ≤6-, and ≤12.5-mm RM and WT differed ($p < 0.001$; **Table 7**). The ≤2-mm RM combination removed 42% more P than the ≤6-mm material, which removed 28% more P than the RM ≤12.5-mm material. While P removed by the finest material was slightly greater for RM and WT (4%, $p < 0.001$), there was no difference in the amount of P removed between RM and WT ≤6-mm materials (**Table 6**). Removed P differed ($p < 0.05$) between RM and WT ≤12.5-mm particle-size treatment, with RM removing 18% more P than WT (**Table 7**). Combining the HC and P-adsorption results, it was suggested that because the ≤6-mm RM and WT treatments could accommodate water from a 25-yr, 30-min storm and adsorbed greater than 56% of the added P, reducing

Table 7. Average and variance of phosphorus (P) removal of red mud (RM) iron filter cake and water treatment (WT) residual as a function of byproduct particle-size fraction.

Byproduct material	Particle-size fraction	P removal		p-value*
		Average	Variance	
	mm	%		
	Red mud iron filter cake			
	<2	98	0.009	<0.001
	<6	56	0.004	<0.001
	<12.5	28	0.221	<0.001
	Water treatment residual			
	<2	94	0.118	0.496
	<6	57	0.052	<0.001
	<12.5	10	0.280	<0.001

*p-value result of comparison across replicates for each byproduct.

the P concentration on average from 6 to 2.5 mg·P·L^{-1}, this material was preferred to the ≤12.5-mm byproduct material for in-field use.

4. Summary and Conclusions

The ≤2-mm/RM and WT byproducts had a large capacity to adsorb additional P with P_{max} values of 25 and 10 mg·P·g^{-1}, respectively. The P_{max} values for RM and WT were compared with other P adsorbents that have been evaluated, and the P_{max} value for RM was greater than 87% of the materials evaluated, and that for WT was greater than 74% of those evaluated. The P_{max} value of any byproduct used is important in determining the viability of a product to be used in terms of capacity and lifespan of on-farm P-removal systems. Further, the HC of both the ≤6- and ≤12.5-mm RM and WT materials were sufficient to transmit runoff resulting from a 25-yr, 30-min storm event.

The P adsorption column study allowed testing of the three particle-size fraction treatments of each byproduct under conditions that were more reflective of field situations around broiler houses than the batch equilibrium method used for the isotherm analyses. The ≤2- and ≤6-mm particle-size fraction treatments removed greater than 50% of P in solution, indicating viability for use in on-farm P-removal systems. Successful design and operation of an on-farm P-removal system utilizing the RM and WT byproducts would be dependent on both the HC and P-removal potential of the selected particle-size fraction. Ultimately, field testing of any design would be necessary to properly evaluate and ensure effectiveness. The optimal particle size from this assessment of RM and WT materials was the ≤6-mm particle-size fraction, which allowed for a combination of sufficient through flow of rainfall generated runoff, while retaining a capacity to adsorb at least 50% of added solution P.

However, the practical implications of this research are that use of these two locally-sourced byproduct materials contained in trays beneath broiler production house ventilation fans, has the potential to trap P exhausted from the fans. Further, as these materials are currently landfilled, their on-farm use could provide a potential cost saving alternative. Finally, the byproduct material, once saturated with P, could become of potential use as a long-term, slow-release fertilizer P source to area pastures.

Acknowledgements

This research was supported by a grant from the USGS 104B program funding.

References

[1] Scavia, D.J., Allan, D., Arend, K.K., Bartell, S., Beletsky, D., Bosch, N.S., Brandt, S.B., Briland, R.D., Daloglu, I., DePinto, J.V., Dolan, D.M., Evans, M.A., Farmer, T.M., Goto, D., Han, H., Hook, T.O., Knight, R., Ludsin, S.A., Mason, D., Michalak, A.M., Richards, R.P., Roberts, J.J., Rucinski, D.K., Rutherford, E., Schwab, D.J., Sesterhenn, T.M., Zhang, H. and Zhou, Y. (2014) Assessing and Addressing the Re-Eutrophication of Lake Erie: Central Basin Hypoxia. *Journal of Great Lakes Research*, **40**, 226-246. http://dx.doi.org/10.1016/j.jglr.2014.02.004

[2] U.S. Environmental Protection Agency (USEPA) (2015) Watershed Assessment, Tracking & Environmental Results. National Probable Sources Contribution to Impairments. http://iaspub.epa.gov/waters10/attains_nation_cy.control#prob_source

[3] Herron, S.L., Brye, K.R., Sharpley, A.N., Miller, D.M. and Daniels, M.B. (2015) Nutrient Composition of Dust Emitted from Poultry Broiler Houses in Northwest Arkansas. *Journal of Environmental Protection*, **6**, 1257-1267. http://www.scirp.org/journal/PaperInformation.aspx?PaperID=61058 http://dx.doi.org/10.4236/jep.2015.611110

[4] Herron, S.L., Sharpley, A.N., Brye, K.R., Miller, D.M., Watkins, S., McCreery, D. and Daniels, M.B. (2016) Determination of Nutrient Concentrations in Simulated Rainfall-Runoff from Poultry House Dust Deposited Adjacent to Exhaust Fans. *Journal of Environmental Protection*, **7**, 27-40. http://dx.doi.org/10.4236/jep.2016.71003

[5] Sharpley, A.N., Beegle, C., Bolster, C., Good, L.W., Joern, B., Ketterings, Q., Lory, J., Mikkelsen, R., Osmond D. and Vadas, P.A. (2012) Phosphorus Indices: Why We Need to Take Stock of How We Are Doing. *Journal of Environmental Quality*, **41**, 1711-1718. http://dx.doi.org/10.2134/jeq2012.0040

[6] Daniel, T.C., Sharpley, A.N. and Lemunyon, J.L. (1998) Agricultural Phosphorus and Eutrophication: A Symposium Overview. *Journal of Environmental Quality*, **27**, 251-257. http://dx.doi.org/10.2134/jeq1998.00472425002700020002x

[7] Sims, J.T. and Kleinman, P.J.A. (2005) Managing Agricultural Phosphorus for Environmental Protection. In: Sims, J.T. and Sharpley, A.N., Eds., *Phosphorus: Agriculture and the Environment*, Am. Soc. Agron. Monograph, American Society of Agronomy, Madison, WI, 1021-1068.

[8] Sharpley, A.N., Daniel, T.C., Gibson, G., Bundy, L., Cabrera, M., Sims T., Stevens, R., Lemunyon, J., Kleinman, P.J.A. and Parry, R. (2006) Best Management Practices to Minimize Agricultural Phosphorus Impacts on Water Quality. USDA-ARS Publication 163. U.S. Government Printing Office, Washington DC.

[9] Faucette, L.B., Sefton, K.A., Sadeghi, A.M. and Rowland, R.A. (2008) Sediment and Phosphorus Removal from Simulated Storm Runoff with Compost Filter Socks and Silt Fence. *Journal of Soil and Water Conservation*, **63**, 257-264. http://dx.doi.org/10.2489/jswc.63.4.257

[10] Faucette, L.B., Cardoso-Gendreau, F.A., Codling, E., Sgdeghi, A.M., Pachepsky, Y.A. and Shelton, D.R. (2008) Storm Water Pollutant Removal Performance of Compost Filter Socks. *Journal of Environmental Quality*, **38**, 1233-1239. http://dx.doi.org/10.2134/jeq2008.0306

[11] Herron, S., Sharpley, A.N., Watkins, S. and Daniels, M. (2012) Poultry Litter Management in the Illinois River Watershed of Arkansas and Oklahoma. Cooperative Extension Service, Division of Agriculture, University of Arkansas. Fact Sheet FSA 9535. 4 p. http://www.uaex.edu/Other_Areas/publications/PDF/FSA-9535.pdf

[12] Vohla, C., Koiv, M., Bavor, H.J., Chazarenc, F. and Mander, U. (2011) Filter Materials for Phosphorus Removal from Wastewater in Treatment Wetlands—A Review. *Ecological Engineering*, **37**, 70-89. http://dx.doi.org/10.1016/j.ecoleng.2009.08.003

[13] Bishop, M.M., Cornwell, D.A., Rolan, A.T. and Bailey, T. (1991) Mechanical Dewatering of Alum Solids and Acidified Solids: An Evaluation. *Journal of American Water Works Association*, **83**, 50-55.

[14] Chardon, W.J., Groenenberg, J.E., Temminghoff, E.J.M. and Koopmans, G.F. (2012) Use of Reactive Materials to Bind Phosphorus. *Journal of Environmental Quality*, **41**, 636-646. http://dx.doi.org/10.2134/jeq2011.0055

[15] Penn, C.J., Bryant, R.B., Kleinman, P.J.A. and Allen, A.L. (2007) Removing Dissolved Phosphorus from Drainage Ditch Water with Phosphorus Sorbing Materials. *Journal of Soil and Water Conservation*, **62**, 269-276.

[16] Oliver, I.W., Grant, C.D. and Murray, R.S. (2011) Assessing Effects of Aerobic and Anaerobic Conditions on Phosphorus Sorption and Retention Capacity of Water Treatment Residuals. *Journal of Environmental Management*, **92**, 960-966. http://dx.doi.org/10.1016/j.jenvman.2010.11.016

[17] Murphy, J. and Riley, J.P. (1962) A Modified Single Solution Method for Determination of Phosphate in Natural Waters. *Analytica Chimica Acta*, **27**, 31-36. http://dx.doi.org/10.1016/S0003-2670(00)88444-5

[18] Syers, J.K., Brownman, M.G., Smillie, G.W. and Corey, R.B. (1973) Phosphate Sorption by Soils Evaluated by the Langmuir Adsorption Equation. *Soil Science Society of America Journal*, **37**, 358-363. http://dx.doi.org/10.2136/sssaj1973.03615995003700030015x

[19] Reynolds, W.D. and Elrick, D.E. (2002) Hydraulic Conductivity of Saturated Soils, Constant Head Method. In: *Methods of Soil Analyses*, *Part* 4, *Physical Methods*, Book Series 5, Soil Science Society of America, Madison, WI, 694-700.

[20] U. S. Department of Commerce (USDC) (1961) Rainfall Frequency Atlas of the United States. Technical Paper 40, U.S. Government Printing Office, Washington DC, 65 p.

[21] Sibrell, P.L., Montgomery, G.A., Ritenour, K.L. and Tucker, T.W. (2009) Removal of Phosphorus from Agricultural Wastewaters Using Adsorption Media Prepared from Acid Mine Drainage Sludge. *Water Research*, **43**, 2240-2250. http://dx.doi.org/10.1016/j.watres.2009.02.010

[22] Sekhon, B.S. and Bhumbla, D.K. (2012) Phosphorus Remediation by Acid Mine Drainage floc and Its Implications for Phosphorus Environmental Indices. *Journal of Soils and Sediments*, **13**, 336-343. http://dx.doi.org/10.1007/s11368-012-0621-y

[23] Bellier, N., Chazarenc, F. and Comeau, Y. (2006) Phosphorus Removal from Wastewater by Mineral Apatite. *Water Research*, **40**, 2965-2971. http://dx.doi.org/10.1016/j.watres.2006.05.016

[24] Fenton, O., Healy, M.G., Rodgers, M. and Huallachain, D.O. (2009) Site-Specific P Absorbency of Ochre from Acid Mine-Drainage near an Abandoned Cu-S Mine in the Avoca-Avonmore Catchment, Ireland. *Clay Minerals*, **44**, 113-123. http://dx.doi.org/10.1180/claymin.2009.044.1.113

[25] Drizo, A., Forget, C., Chapuis, R.P. and Comeau, Y. (2006) Phosphorus Removal by Electric Arc Furnace Steel Slag and Serpentinite. *Water Research*, **40**, 1547-1554. http://dx.doi.org/10.1016/j.watres.2006.02.001

[26] Mann, R.A. (1997) Phosphorus Adsorption and Desorption Characteristics of Constructed Wetland Gravels and Steelworks by-Products. *Australian Journal of Soil Research*, **35**, 375-384. http://dx.doi.org/10.1071/S96041

[27] Yan, J., Kirk, D.W., Jia, C.Q. and Liu, X. (2007) Sorption of Aqueous Phosphorus onto Bituminous and Lignitous Coal Ashes. *Journal of Hazardous Materials*, **148**, 395-401. http://dx.doi.org/10.1016/j.jhazmat.2007.02.055

[28] Prochaska, C.A. and Zouboulis, A.I. (2006) Removal of Phosphates by Pilot Vertical-Flow Constructed Wetlands Using a Mixture of Sand and Dolomite as Substrate. *Ecological Engineering*, **26**, 293-303. http://dx.doi.org/10.1016/j.ecoleng.2005.10.009

[29] Pant, H.K., Reddy, K.R. and Lemon, E. (2001) Phosphorus Retention Capacity of Root Bed Media of Subsurface Flow Constructed Wetlands. *Ecological Engineering*, **17**, 345-355. http://dx.doi.org/10.1016/S0925-8574(00)00134-8

[30] Chen, J., Kong, H., Wu, W., Chen, X., Zhang, D. and Sun, Z. (2007) Phosphate Immobilization from Aqueous Solution by Fly Ashes in Relation to Their Composition. *Journal of Hazardous Materials*, **139**, 293-300. http://dx.doi.org/10.1016/j.jhazmat.2006.06.034

[31] Dobbie, K.E., Heal, K.V., Aumonier, J., Smith, K.A., Johnston, A. and Younger, P.L. (2009) Evaluation of Iron Ochre from Mine Drainage Treatment for Removal of Phosphorus from Wastewater. *Chemosphere*, **75**, 795-800. http://dx.doi.org/10.1016/j.chemosphere.2008.12.049

[32] Drizo, A. (1998) Phosphate and Ammonium Removal from Waste Water, Using Constructed Wetland Systems. PhD Thesis, University of Edinburgh, Edinburgh.

[33] Heal, K.V., Smith, K.A., Younger, P.L., McHaffie, H. and Batty, L.C. (2004) Removing Phosphorus from Sewage Effluent and Agricultural Runoff Using Recovered Ochre. In Valsami-Jones, E., Ed., *Phosphorus in Environmental Technology: Principles and Applications*, IWA Publishing, London, 321-334.

[34] Arias, C.A., Del Bubba, M. and Brix, H. (2001) Phosphorus Removal by Sands for Use as Media in Subsurface Flow Constructed Reed Beds. *Water Research*, **35**, 1159-1168. http://dx.doi.org/10.1016/S0043-1354(00)00368-7

[35] Roseth, R. (2000) Shell Sand a New Filter Medium for Constructed Wetlands and Wastewater Treatment. *Journal of Environmental Science and Health Part A*, **35**, 1335-1355. http://dx.doi.org/10.1080/10934520009377039

[36] Sovik, A.K. and Klove, B. (2005) Phosphorus Retention Processes in Shell sand Filter Systems Treating Municipal Wastewater. *Ecological Engineering*, **25**, 168-182. http://dx.doi.org/10.1016/j.ecoleng.2005.04.007

[37] Adam, K., Krogstad, T., Vrale, L., Sovik, A.K. and Jenssen, P.D. (2007) Phosphorus Retention in the Filter Materials Shells and Filtralite P®—Batch and Column Experiment with Synthetic P Solution and Secondary Wastewater. *Ecological Engineering*, **29**, 200-208. http://dx.doi.org/10.1016/j.ecoleng.2006.09.021

[38] Wei, X., Viadero Jr., R.C. and Bhojappa, S. (2008) Phosphorus Removal by Acid Mine Drainage Sludge from Secondary Effluents of Municipal Wastewater Treatment Plants. *Water Research*, **42**, 3275-3284. http://dx.doi.org/10.1016/j.watres.2008.04.005

[39] Sakadevan, K. and Bavor, H.J. (1998) Phosphate Adsorption Characteristics of Soils, Slags and Zeolite to Be Used as Substrates in Constructed Wetland Systems. *Water Research*, **32**, 393-399. http://dx.doi.org/10.1016/S0043-1354(97)00271-6

3

Form Distribution Characteristics of Nitrogen in a Reservoir as Drinking Water Source

Lahua Jin, Chen Zhao, Jiayi Wang

School of Environment, Jinan University, Guangzhou, China
Email: profjin@163.com

Abstract

Based on field detected water quality data, the distribution characteristics of different forms of nitrogen in a reservoir as drinking water source in Dongguan, which locates at the Pearl River Delta of China, have been analyzed in order to provide theoretical bases for prevention and reduction of eutrophication. The analyzed results show that nitrogen forms in the influent area of the reservoir are given priority to ammonia nitrogen and nitrate nitrogen, whose proportion is more than 45% respectively, and this is probably caused by the pollution of inflow water quality; but in the effluent area, the forms are given priority to nitrate nitrogen, whose proportion is as high as 96% and above; also the proportion of ammonia nitrogen drops by more than 80% during the process from the influent area to the effluent area, and this shows that the natural process of nitrification and denitrification can be well accomplished in the reservoir. We recommend here that to reduce the input amount of ammonia nitrogen and organic nitrogen into the reservoir is the most efficient way to prevent or mitigate eutrophication of water body.

Keywords

Nitrogen Form, Reservoir for Drinking Water Source, Eutrophication, Dissolved Organic Nitrogen

1. Introduction

Nitrogen pollution in a reservoir not only can promote the degree of water body eutrophication, but also will generate disinfection by-products of nitrogen (N-DBPs) during the process of chlorine disinfection when the water is pumped to prepare for dinking. The N-DBPs are carcinogenic and most harmful to human health. In order to control this risk, it is necessary to make clear distribution regulation of nitrogen forms and their variation in a reservoir.

But up-to-date, researches on nitrogen form characteristics in a reservoir are still few. Xiaoli Ge *et al.* investigated geochemical characteristics of water body in Reservoir Miyun in Beijing during 2001, and found out that TN/TP ratio was as high as 38.4:1 and pointed out that phosphorus is the major nutrient element limiting biological productivity [1]. Yeqing Feng *et al.* investigated the mean concentrations of TN, TP, chlorophyll-a and SD in Reservoir Hongfeng located on the Guizhou plateau in southwest China in 2010 as 1.83 mg/L, 0.034 mg/L, 0.0127 mg/L and 2.3 m respectively, which shows that the water body in the reservoir was in meso-eutrophic state [2]. Yuejuan Lin analyzed and assessed the characteristics of sediment cores and their overlaying water quality. The results show that concentration of total nitrogen is relatively low in both summer and winter but much high in both spring and autumn [3]. Yunxian Dong *et al.* established 9 sampling plots on 3 sections in Lake Chenghai located in Yunnan of China by global positioning system to study the speciation and distribution of nitrogen and phosphorus, and their relationships with chlorophyll-a, and the results indicated that the total nitrogen, total phosphorus and total chlorophyll-a in the lake water were 0.773 mg/L, 0.046 mg/L and 0.024 mg/L respectively, and the majority of the total nitrogen was dissolved nitrogen in which dissolved organic nitrogen was predominant, while dissolved inorganic phosphorus consisted of majority of the total phosphorus [4]. Fengjiao Wang analyzed the geochemical behavior of nitrogen, phosphorus and other biogenic elements in the overlaying water and sediment of the Hulun Lake in arid areas within the Mongolian Plateau, and found out that the main existing form of nitrogen is the organic nitrogen in the lake [5]. Honelei Yu (2013) analyzed the distribution and transformation of DON in both landscape water and effluent water from different treatment processes in municipal sewage treatment plants, and concluded that there was seasonal variation in the concentrations of DON in the landscape water, and there was significant difference between the north and south part in different months. And DON in landscape water is the potential nutrition of algae, and it is remarkable in the process of eutrophication control [6]. So far, the distribution and its variation of nitrogen forms in a reservoir as drinking water source are still not clear and need to be further deeply studied. Thus, we have taken Reservoir TS, which locates at Dongguan of Guangdong Province in China, as an example to analyze the distribution characteristics and its spatial and temporal variation of nitrogen forms in reservoir.

2. Survey of the Reservoir

2.1. Basic Condition

Reservoir TS is located in southeast of Dongguan city. It has catchment area of 100 km^2 and total capacity of 65.2 million m^3, as shown in **Figure 1**. It was classified as one of the backup drinking water sources for the city in 2006. Huanyan Huang *et al.* indicated after analyzed the detected water quality data that the reservoir is in a state of slight pollution [7], and Jiao Han discussed the environmental capacity of the reservoir [8].

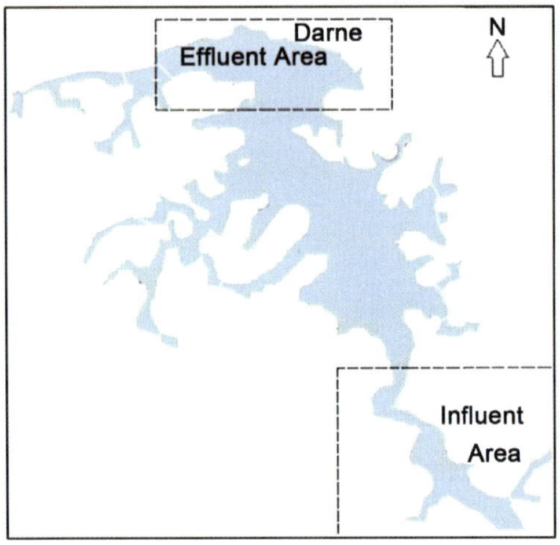

Figure 1. Plan of reservoir TS.

2.2. Survey Result

During the period from April 2014 to January 2016, we sampled water quality and nutritional level of the reservoir. According to the water quality requirement of *China National Environmental Quality Standards for Surface Water* (GB3838-2002), the single factor standard indexes of the water quality have been calculated out as shown in **Table 1** based on the detected data. It can be seen from the table that there exist five indexes exceeded the national standard, *i.e.*, COD_{Mn}, COD_{cr}, ammonia nitrogen, BOD_5 and total phosphorus. Thus, it has been shown from this phenomenon that the reservoir has been obviously contaminated by organic and nutrient pollutants.

Based on the monitored data of water body nutritional level, the comprehensive nutritional status index of the reservoir has been calculated out as shown in **Figure 2** according to the improved comprehensive nutrition state index calculation method provided by Rui Zhang *et al.* [9]. It has been shown from **Figure 2** that the reservoir is at meso-eutrophic status, but its nutritional level has a decrease trend. But the degree of comprehensive nutrition in the effluent area is lower than that in the influent area of the reservoir. This shows that the reservoir has good self-purification ability because nutrient pollutants can be degraded obviously from influent area to effluent area.

3. Nitrogen Form Distribution Characteristics

Nitrogen in water can be mainly divided into inorganic nitrogen and organic nitrogen. Usually, inorganic nitrogen includes ammonia nitrogen and nitric nitrogen, in which the ammonia nitrogen includes free ammonia nitrogen $(NH_3 - N)$ and ammonium nitrogen $(NH_4^+ - N)$, and the nitric nitrogen includes nitrate nitrogen $(NO_3 - N)$ and nitrite nitrogen $(NO_2 - N)$. But organic nitrogen in water contains a lot of organic compounds, such as urea, amino acid, protein, nucleic acid, uric acid, fatty amine, organic base, amino sugar and so on. The concentration and proportion of different forms of nitrogen in water are often determined by specific circumstances of the water body.

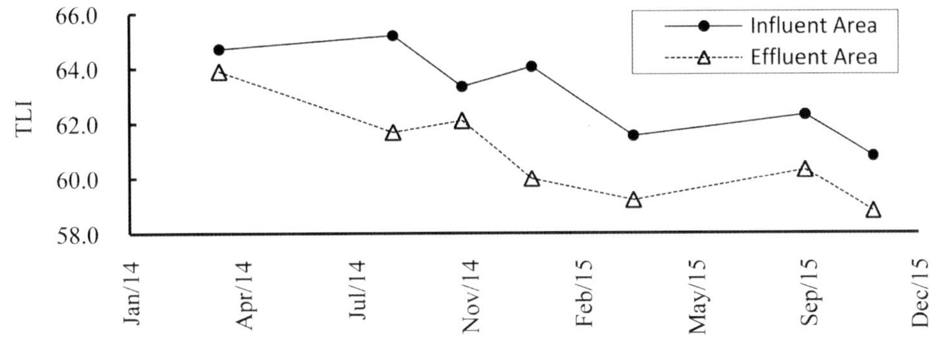

Figure 2. Variation of the comprehensive nutritional status index TLI of the reservoir.

Table 1. Present status of water quality standard indexes of the reservoir.

Item	Standard Index			Status
	Influent area	Effluent area	Average	
pH	0.085	0.098	0.091	Qualified
DO	0.273	0.204	0.239	Qualified
COD_{Mn}	1.080	0.963	1.022	Exceed standard
COD_{cr}	1.020	1.370	1.195	Exceed standard
Ammonia nitrogen	2.768	1.498	2.133	Exceed standard
BOD_5	1.265	1.690	1.478	Exceed standard
TP	2.960	2.240	2.600	Exceed standard
Petroleum class	0.800	0.800	0.800	Qualified
Anionic surface-active agent	0.150	0.150	0.150	Qualified
Volatile phenol	0.100	0.100	0.100	Qualified
Sulfide	0.015	0.015	0.015	Qualified
Cyanide	0.003	0.003	0.003	Qualified
Zinc	0.010	0.008	0.009	Qualified
Copper	0.004	0.003	0.004	Qualified
Lead	0.001	0.001	0.001	Qualified
Cadmium	0.006	0.006	0.006	Qualified
Selenium	0.010	0.010	0.010	Qualified
Mercury	0.050	0.050	0.050	Qualified
Arsenic	0.023	0.024	0.024	Qualified
Chromium of six valence	0.040	0.040	0.040	Qualified
Fluoride	0.672	0.538	0.605	Qualified

3.1. Proportion of Different Forms of Nitrogen

The proportion of different forms of nitrogen in the reservoir has been analyzed base on the detected data of water quality, and the result shows that inorganic nitrogen occupies most proportion of total nitrogen, and its proportion is as high as over 92% in the influent area or over 84% in the effluent area at any time, but the organic nitrogen only occupies below 8% in the influent area or below16% in the effluent area at any time, as shown in **Figure 3**.

The inorganic nitrogen in the reservoir is found to be composed of ammonia nitrogen and nitric nitrogen, and their average proportions during the detecting period are 49.9% and 50.1% in the influent area, and 48.5% and 51.5% in the effluent area respectively, as shown in **Table 2**.

The nitric nitrogen in the reservoir is also found to be composed of about 96% - 99% of nitrate nitrogen and 1% - 4% of nitrite nitrogen, as shown in **Table 3**.

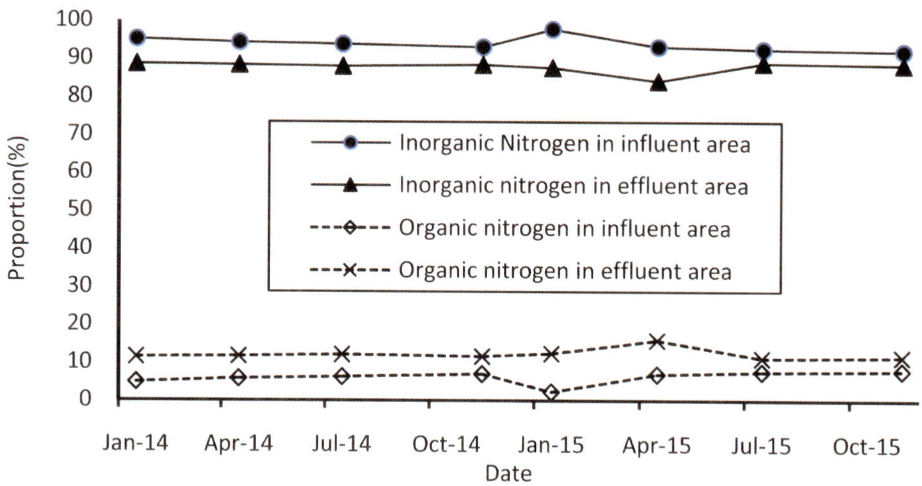

Figure 3. Variation of proportion of inorganic and organic nitrogen in the reservoir.

Table 2. Composition proportion of inorganic nitrogen in the reservoir.

Nitrogen form		Proportion (%)								
		Jan-14	Apr-14	July-14	Nov-14	Jan-15	Apr-15	July-15	Nov-15	Average
Ammonia Nitrogen	Influent	52	51	51	51	49	48	48	49	49.9
	Effluent	51	50	49	49	48	47	47	47	48.5
Nitric Nitrogen	Influent	48	49	49	49	51	52	52	51	50.1
	Effluent	49	50	51	51	52	53	53	53	51.5

Table 3. Composition proportion of nitric nitrogen in the reservoir.

Nitrogen form		Proportion (%)								
		Jan-14	Apr-14	July-14	Nov-14	Jan-15	Apr-15	July-15	Nov-15	Average
Nitrate Nitrogen	Influent	98	98	98	98	97	96	96	96	97.1
	Effluent	99	99	99	98	97	96	96	96	97.5
Nitrite Nitrogen	Influent	2	2	2	2	3	4	4	4	2.9
	Effluent	1	1	1	2	3	4	4	4	2.5

It can be concluded from the above analyses that the proportion of different forms of nitrogen in the reservoir are 45.0% of nitrate nitrogen, 44.7% of ammonia nitrogen, 9.1% of organic nitrogen and 1.2% of nitrite nitrogen.

3.2. Spatial Distribution of Different Forms of Nitrogen

The spatial distribution of inorganic nitrogen and organic nitrogen has obvious differences in the reservoir, *i.e.*, during the process from the influent area to the effluent area, the inorganic nitrogen will degrade about 6.1%, but the organic nitrogen will increase about 6.1%, as shown in **Figure 3**.

The degradation of inorganic nitrogen may be caused by the reaction of its nitrification and denitrification. This phenomenon demonstrates that the water quality is getting improved and the self-purification ability of water is strong in the reservoir.

4. Conclusions

Nitrogen forms in a reservoir as drinking water source behave as over 84% of inorganic nitrogen and below 16% of organic nitrogen. The inorganic nitrogen is composed of ammonia nitrogen and nitric nitrogen, in which the nitric nitrogen is mostly composed by nitrate nitrogen.

The proportion of inorganic nitrogen in the influent area is some higher than those in the effluent area, but the proportion of organic nitrogen is little lower in the influent area than those in the effluent area.

Reservoirs as drinking water source may have essential capability of self-purification to degrade nutrient pollution.

Acknowledgements

This work has been supported by Guangdong provincial scientific technology project fund (No. 2015A020215015).

References

[1] Ge, X.L., Liu, X.D., Pan, X.C., Li, Q. and Luo, S.-G. (2003) The Geochemical Characteristics of Water Bodyin Miyun Reservoir. *Rock and Mineral Analysis*, **22**, 44-48.

[2] Feng, Y.Q., Xia, P.H., Zhang, M.S., Li, C.X., Lin, T., Ma, J.R., Long, J. and Yang, J.J. (2011) Analysis on Eutrophication Features of Hongfeng Reservoir on Guizhou Plateau. *Journal of Guizhou Normal University (Natural Sciences)*, **29**, 29-35.

[3] Lin, Y.J. (2005) Distribution of Nitrogen and Phosphorus in Sediment Core and Overlay Water of Donghu Lake. Master's Degree Thesis, Wuhan University, Wuhan.

[4] Dong, Y.X., Hong, X.H., Tan, Z.W., Zhu, X. and Li, Y.Q. (2012) Distribution of Nitrogen and Phosphorus and Their Relationships with Chlorophylla in Lake Chenghai on Plateau. *Journal of Ecology and Environmental Sciences*, **21**, 333-337.

[5] Wang, F.J. (2010) Geochemistry Characteristics of Nitrogen and Phosphorus of the Hulun Lake. Master's Degree Thesis, Inner Mongolia University, Huhhot.

[6] Yu, H.L. (2013) Study on the Characteristics of Dissolved Organic Nitrogen in Wastewater and Landscape Water. Master's Degree Thesis, Anhui Normal University, Wuhu.

[7] Huang, H.Y., Zeng, R., Li, X.W., Xia, H.Z. and Liu, H.R. (2007) Fuzzy Mathematics Comprehensive Evaluation of Water Environment Quality of Reservoir Tongsha. *Chinese Journal of Industrial Water Usage & Wastewater*, **42**, 42-44.

[8] Han, J. (2013) Analysis on Water Environmental Capacity of the Tongsha Reservoir. *Chinese Journal of Henan Science and Technology*, No. 4, 180-181.

[9] Zhang, R., Gao, L.M., Xi, B.D., Su, J., Huo, S.L., *et al.* (2014) Improved TLI Index Method and Its Application in Nutritional States Evaluation in Chaohu Lake. *Chinese Journal of Environmental Engineering*, **7**, 2127-2132.

4

Study on the Effect of Different Modified Zeolite to Phosphorus Activation in Red Soil

Jing Shi[1,2], Naiming Zhang[1,2*], Chenguang Zhang[1,2], Boxian Wei[1,2]

[1]College of Resource and Environment, Yunnan Agricultural University, Kunming, China
[2]Soil Fertilization and Pollution Remediation Engineering Laboratory of Yunnan Province, Kunming, China
Email: 383110966@qq.com, *zhangnaiming@sina.com

Abstract

The activation of soil-fixed phosphorus is a long-term concern in soil science. In order to enhance the activation effect to soil-fixed phosphorus in red soil, different modified zeolite, through physical, chemical method and the associated both physics and chemistry, was used in this paper. The results showed as following: 1) the activated ability to soil-fixed phosphorus in red soil for modified zeolite was higher than the common one signally. Order of the ability for all kinds of modification zeolite in absolute activation amount and activation rate was: Modified in high temperature ammonium saturated zeolite > Ammonium saturated zeolite > Modified in high temperature zeolite > Common zeolite. 2) The increasing rate of modified phosphorus in high temperature ammonium saturated zeolite and ammonium saturated zeolite had a decline trend with time postponed while that for modified phosphorus in high temperature zeolite and routine zeolite was rising gradually with time postponed. For ammonium saturated zeolite, the increasing rate of phosphorus in low available phosphorus content red soil was significantly faster than that in red soil with the higher content of available phosphorus. 3) The activation rate for the same modified zeolite showed difference in different phosphorus level soil. The order is: $SP_{high} > SP_{medium} > SP_{low}$.

Keywords

Activation, Phosphorus, Red Soil, Zeolite

1. Introduction

Phosphorus is one of the necessary nutritional elements during plant growth and also a limiting factor of water eutrophication [1]. Red soil is the main arable soil in China and its phosphorus content is abundant, but the content of available phosphorus is generally low. Though the phosphorus needs for the crop can be met by using phosphorus fertilization but it is still not an effective way to solve the problem. For one reason, the seasonal crop utilization of phosphate is only 5% - 10% [2] [3]. Besides, most of the available P will be fixed in the soil in the form of fixed phosphorus. For one thing, phosphorus fertilizer is overused in order to ensure the yield and food security. For another, there exist different levels of eutrophication in over 80% of lakes and reservoirs [4] [5] [6]. Phosphate application must be reduced in order to protect the environment and prevent the eutrophication. Meanwhile, phosphate mineral is nonrenewable resource. Considering these factors, to make full use of the fixed state of phosphorus in the soil, in order to balance food production, water environment protection and phosphate mineral resource, has become an inevitable choice. Therefore the research of cost-effective activation method of the soil fixed phosphorus is meaningful.

Studies showed that the accumulation of insoluble phosphate in the soil can be released by the zeolite and therefore as a source of supply of plant available phosphorus [7] [8] [9] [10]. However, the existed methods were not ideal to the activation effect on soil fixed phosphorus. In order to improve the effect of phosphorus activation using zeolite further, the physical method of high-temperature modification, saturated NH_4^+ chemical methods and high-temperature modification with saturated NH_4^+ combination of physical and chemical methods were used in this research. Together with solubilizing simulation experiments in three phosphorus levels, it focused on finding the method that can be applied in fixed phosphorus activation of red soil.

2. Materials and Methods

2.1. Tested Soil and Regional Profiles

Soil samples used in this study were taken from Yunlong Reservoir farmland soil which is the largest urban centralized drinking water source of Kunming. Sampling depth was 5 - 20 cm topsoil. The soil samples were air-dried and ground through a sieve of 1mm aperture and stored in a sealed bag for use. The basic properties of the soils are shown in **Table 1**.

2.2. Design of Experiments

2.2.1. Experiment Materials

Natural zeolite is an aluminum silicate mineral with basic metal frame shape. As it has

Table 1. Experimental soils' basic properties and contents of phosphorus.

Sample Numbers	Soil types	Available phosphorus ($mg \cdot kg^{-1}$)	Total phosphorus ($g \cdot kg^{-1}$)	Organic matter ($g \cdot kg^{-1}$)	pH
SP_{high}	Red soil	42.27	1.217	22.70	5.60
SP_{medium}	Red soil	22.45	0.985	34.47	5.75
SP_{low}	Red soil	9.07	1.023	23.82	5.83

many internal cavities and channels, it is strong in cation exchange and selective adsorption performance, as well as its rich adsorption capacity. Zeolite used in this study collected from Yunnan and its main components are clinoptilolite. Cation exchange capacity is 89.3cmol/kg. Molecular formula is (SiO_2) (Al_2O_3). Its weight is 218.247 and chemical formula is $(CaNa_2)$ $Al_2Si_7O{\cdot}6H_2O$.

2.2.2. The Experimental Design

This experiment set up 4 treatments with 3 replications. 4 treatments are: Treatment 1 was common zeolite treatment; Treatment 2 was high temperature modified zeolite; Treatment 3 was ammonium saturated zeolite and Treatment 4 was high-temperature modification of ammonium saturated zeolite. The proportion ratio of modified zeolite is 1:100 and three red soil samples with SP_{high}, the SP_{medium} and SP_{low} were added according to this ratio. Soil sample was taken every 10 d with a total of 4 times. The available phosphorus content was determined every time, the absolute amount of the activation of phosphorus was calculated by subtracting the available phosphorus content in CK treatment from the available phosphorus content of each zeolite activation treatment; the increasing rate of available phosphorus was calculated by dividing increasing amount of absolute available phosphorus by available phosphorus content in CK then multiplied by 100%. By subtracting the increasing rate of previous available phosphorus from the last available phosphorus, calculating the dynamic increase rate of available Phosphorus, the activation rate of Phosphorus was calculated from dividing the increase rate of absolute available phosphorus by total phosphorus in CK soil then multiplied by 100%.

2.3. The Method of Determining Index

Soil available phosphorus was measured by sodium bicarbonate extraction-molybdenum antimony anti-colorimetric determination method. Total phosphorus in soil use NaOH-molten molybdenum, antimony anti-colorimetric determination method. pH value was tested by the precision pH meter with PHS-3C; the soil organic matter uses external heating potassium dichromate oxidation method for determination.

3. Results and Discussion

3.1. Effect and Variance Analysis of Different Modification Zeolite to Activate Red Soil Phosphorus

3.1.1. Effect of Different Modification Zeolite to Activate Red Soil Phosphorus

It showed that the absolute increment of available phosphorus with high-temperature modification of ammonium saturated zeolite was the highest between the four modified zeolites (**Table 2**). The absolute increments of available phosphorus were 23.48 $mg{\cdot}kg^{-1}$, 15.10 $mg{\cdot}kg^{-1}$ and 7.92 $mg{\cdot}kg^{-1}$. Solubilizing effect of the ammonium saturated zeolite was also evident, the absolute increments of its available phosphorus were 20.30 $mg{\cdot}kg^{-1}$, 12.35 $mg{\cdot}kg^{-1}$ and 6.13 $mg{\cdot}kg^{-1}$. Solubilizing effect of the high-temperature modified zeolite was better than the zeolite, and the absolute increments of its available phosphorus were 7.20 $mg{\cdot}kg^{-1}$, 6.90 $mg{\cdot}kg^{-1}$ and 3.49 $mg{\cdot}kg^{-1}$. The absolute increments

Table 2. The absolute activation amount of different modification zeolite to activate red soil phosphorus.

Modification treatment	SP_h	SP_m	SP_l
	Absolute increments of Available phosphorus($mg \cdot kg^{-1}$) (average ± Standard deviation)		
Ordinary zeolite	1.10 ± 0.2c	2.11 ± 1.5c	0.35 ± 0.4c
Modified zeolite in high temperature	7.20 ± 2.0b	6.90 ± 2.6b	3.49 ± 2.2bc
Ammonium saturated zeolite	20.30 ± 3.1a	12.35 ± 2.6a	6.13 ± 3.2ab
High temperature modified ammonium saturated zeolite	23.48 ± 1.6a	15.10 ± 2.9a	7.92 ± 4.1a

of ordinary zeolite were 1.10 $mg \cdot kg^{-1}$, 2.11 $mg \cdot kg^{-1}$ and 0.35 $mg \cdot kg^{-1}$. It follows that the ordinary zeolite could activate the fixed state phosphorus in the soil but its activation effect was not obvious.

Overall, the combination of Zeolite and saturated NH_4^+ can significantly improve the solubilizing phosphorus amount of zeolite. The main reason is to add saturated NH_4^+ can increase the cation exchange capacity of zeolite. So that the exchange absorption amount of zeolite to free state metal cations in the soil (of Fe^{2+}, of Al^{3+}, of Ca^{2+}) was effectively increased and the fixation to phosphate ion was reduced [11]. Meanwhile soluble phosphorus was released through the cation exchange interaction of Ca^{2+} from insoluble calcium phosphate [12]. The exchange capacity order of different absorption ion to Ca^{2+} was as follows [13]: $H^+ > Na^+ >$ of $NH_4^+ > K^+ > Mg^{2+} > Ba^{2+}$. It follows that different cation has great difference in selective exchange to Ca^{2+}. The main reason that the high temperature modified zeolite has better solubilizing phosphorus effect than ordinary zeolite is the adsorption in the zeolite voids with a certain amount of organic matter. The solubilizing effect of the zeolite is hindered by the relatively stable nature of organic matter, for there is no reaction with insoluble phosphate by the cation exchange role, and also the occupation of the organic matter to zeolite in adsorption capacity. High-temperature burning can effectively remove the organic matter adsorbed on the zeolite voids and thus increase the adsorption ability and adsorption capacity of the zeolite, and its phosphate solubilizing capacity will be significantly better than the zeolite without the high-temperature burning. The solubilizing effect of untreated zeolite is not obvious, it has the reasons: Firstly, the tested soil is red soil with strong acid, and this soil has secondary phosphorus fixation which make the secondary fixation of activated phosphate ion [14]; Secondly, low application amount of zeolite make its cation exchange and adsorption capacity relatively smaller, which is similar with the research result of Li [10].

3.1.2. Variance of Different Modification Zeolite to Activate Red Soil Phosphorus

It shows in **Figure 1(a)** that there was no significant difference between high-temperature modified of ammonium saturated zeolite and ammonium saturated zeolite in solubilizing phosphorus effect, but their solubilizing phosphorus effect was highly

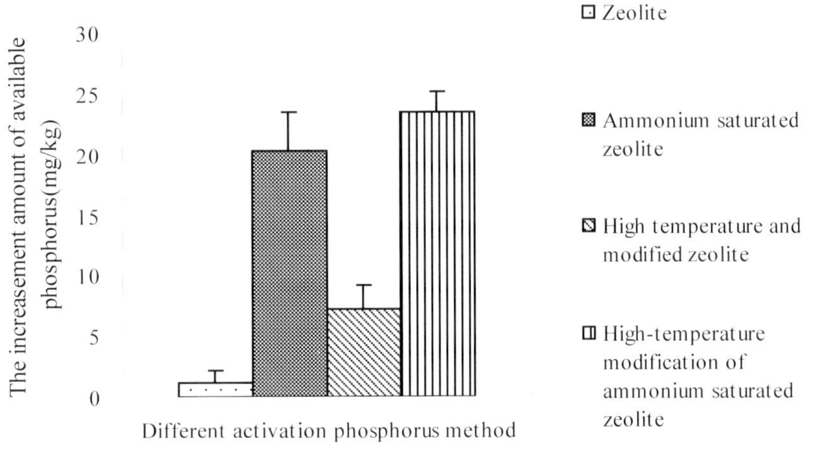

(a)

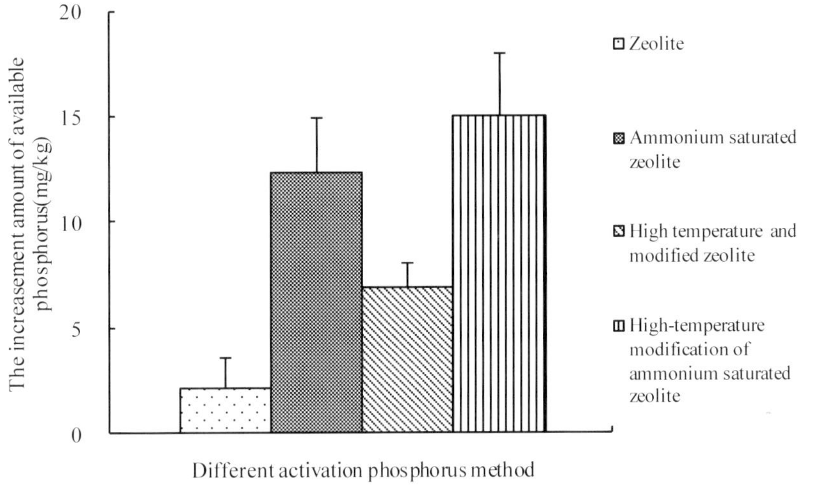

(b)

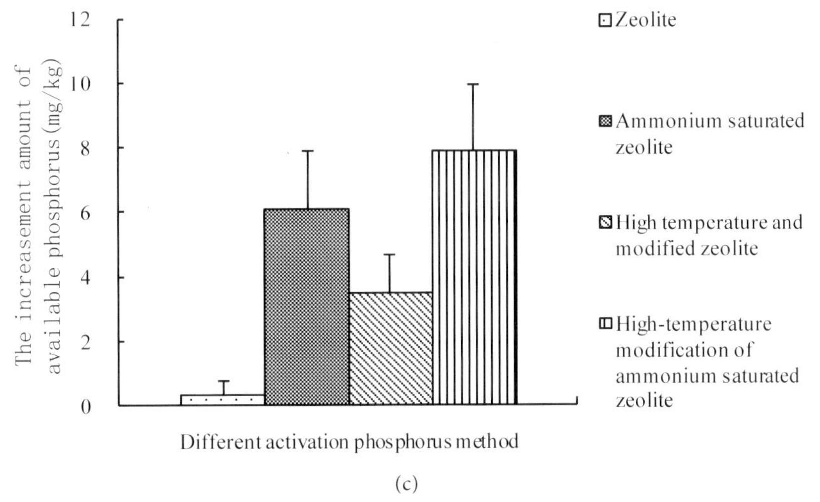

(c)

Figure 1. (a) The difference of (PS high) phosphorus activation effect in different modification treatment zeolite; (b) The difference of (PS medium) phosphorus activation effect in different modification treatment zeolite; (c) The difference of (PS low) phosphorus activation effect in different modification treatment zeolite.

different with the high-temperature-modified zeolite; Solubilizing phosphorus effect of the high-temperature modified zeolite was also significantly different with ordinary zeolite. Regarding the red soil with high SP, the solubilizing effect order of zeolite from four different treatments are: high-temperature modified ammonium saturated zeolite > ammonium saturated zeolite > high temperature modified zeolite > zeolite. **Figure 1(b)** shows that High-temperature modified ammonium saturated zeolite has no big difference with ammonium saturated zeolite about solubilizing effect; their solubilizing effects are highly different with high temperature modified zeolite. It also has significant difference between high-temperature modified zeolites and zeolite. The solubilizing effect of Zeolite for the red soil with medium level SP was: high temperature modification of ammonium saturated zeolite > ammonium saturated zeolite > high temperature modified zeolite > zeolite. **Figure 1(c)** shows that there is no significant difference between high-temperature modified ammonium saturated zeolite and ammonium saturated zeolite, but it has no common with high temperature modified zeolite in solubilizing effect; The solubilizing effect of the ammonium saturated zeolite has no great difference with high temperature modified zeolite but different with Zeolite; The difference was not vital between high temperature modified zeolite and zeolite. The solubilizing effect order of four treatments with low SP red soil was: high-temperature modified ammonium saturated zeolite > ammonium saturated zeolite > high temperature modified zeolite > zeolite.

3.2. Influence of Different Modified Zeolite on the Dynamic Increasing Rate of Available Phosphorus in Red Soil

It can be seen that in the same period, the increasing rates of available phosphorus in red soil are different for the zeolite with different treatments (**Figures 2(a)-(c)**). The dynamic increasing rate of available phosphorus from 3 soil samples in the 0 - 10 d period of high-temperature modified ammonium saturated zeolite were 19.61%, 22.50% and 26.53%, which were higher than the other three periods. Within the 0 - 10 d of ammonium saturated zeolite, the dynamic increasing rate of available phosphorus from 3 soil samples were 17.17%, 18.06% and 30.34%, which were significantly higher than the results from other three periods. And the dynamic increasing rate of available phosphorus with these two treated zeolite showed a descending trends. In addition, in the 0 - 10 d for high temperature modified zeolite, the dynamic increasing rate of available phosphorus from three soil samples was 2.35%, 3.54% and 5.15%, which were lower than the results of other three periods, and their dynamic increasing rate of available phosphorus showed ascending trends. And the zeolite also has the similar pattern.

The reason that the dynamic increasing rate of available phosphorus from high-temperature modified ammonium saturated zeolite and ammonium saturated zeolite showing descending trends is that zeolite absorbed saturated NH_4^+ in extremely rich content, which enhance the cation exchange ability and amount of zeolite, and the activated phosphorus amount was also raised through cation exchange, but over time,

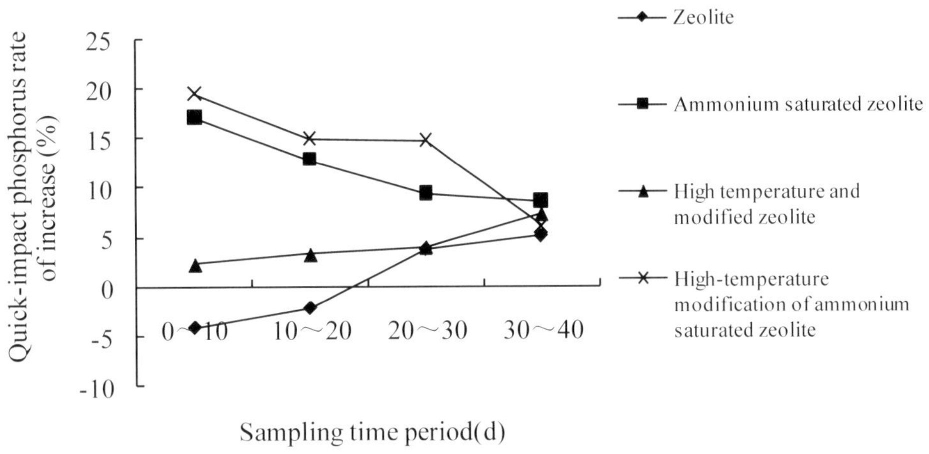

(a)

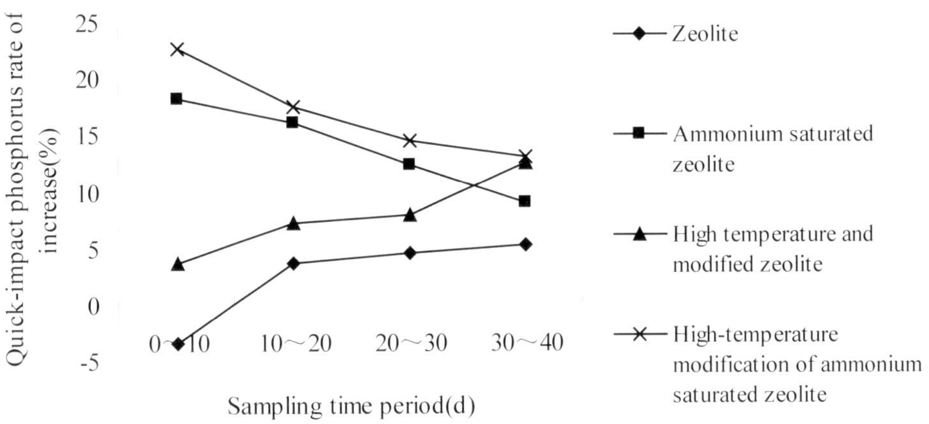

(b)

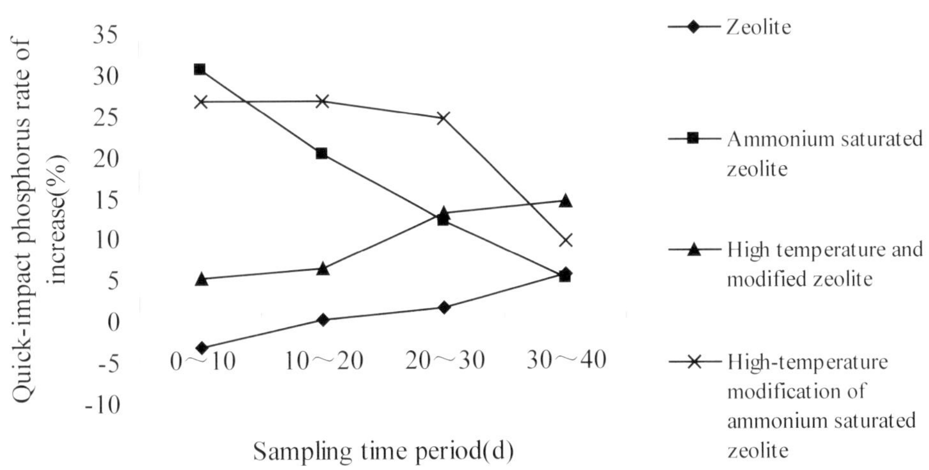

(c)

Figure 2. (a) (PS high) dynamic increasing rate of phosphorus in different modification treatment zeolite; (b) (PS medium) dynamic increasing rate of phosphorus in different modification treatment zeolite; (c) (PS low) dynamic increasing rate of phosphorus in different modification treatment zeolite.

gradually the saturated NH_4^+ was consumed and cation exchange amount was reduced, so that the dynamic increasing rate of available P was declined. The dynamic increasing rate of available phosphorus of high temperature modified zeolite and the ordinary zeolite showing ascending trends is because that their solubilizing mechanism was a slow reaction process, and at the beginning zeolite need to absorb the ions K^+ and NH^{4+} that has cation exchange ability in soil, then by cation exchanges to react with the metal cation of insoluble phosphorus in soil, the phosphate ion will be released slowly [15]. It is just as the existence of this time difference that there have been followed by an upward trend of the dynamic increasing rate of available phosphorus, and also precisely as this phenomenon, the zeolite cannot play the role of activating soil phosphorus and even make negative effect. This could be the reason in the early stage that zeolite showed negative value.

3.3. Effect of Different Modified Zeolite on the Activation Rate of Phosphorus in Red Soil

Figures 3(a)-(c) showed that the different treated zeolite has variable activation phosphorus rates to the same soil sample. The modified ammonium saturated zeolite has the highest activation rate of phosphorus, and the phosphorus activation rates in three soil samples with high SP, medium SP, and low SP are 1.93%, 1.53% and 0.77%. It follows from that the phosphorus activated mode of high-temperature modified ammonium saturated zeolite has the best effect in high SP. It was seen in **Figure 3(a)** that there was no significant difference of phosphorus activation rate between high-temperature modified ammonium saturated zeolite and ammonium saturated zeolite, the difference are significant between ammonium saturated zeolite and high-temperature modified zeolite, and also between high temperature modified zeolite and ordinary zeolite. It was seen in **Figure 3(b)** that there was no significant difference of phosphorus activation rate between high-temperature modified ammonium saturated zeolite and ammonium saturated zeolite, and big difference between ammonium saturated zeolite and high-temperature modified zeolite, and also big difference between high temperature modified zeolite and ordinary zeolite. **Figure 3(c)** showed that high-temperature modified ammonium saturated zeolite had no difference with ammonium saturated zeolite, but it had vital difference with high temperature modified zeolite; Ammonium saturated zeolite had no difference with high temperature modified zeolite, but it had highly difference with ordinary zeolite; and also big difference between high temperature modified zeolite and ordinary zeolite.

As the different mechanisms of activating phosphorus between different processed zeolite, there was a big difference for phosphorus activation effect of the same soil. High-temperature modified ammonium saturated zeolite has the best activating phosphorus effect in the four processed zeolite because the soil organic matter absorbed in the zeolite voids was removed by high temperature, and the absorption capacity of zeolite was enlarged and cation exchange capacity of zeolite was also increased through combination with saturated NH_4^+. This combined physical and chemical method can

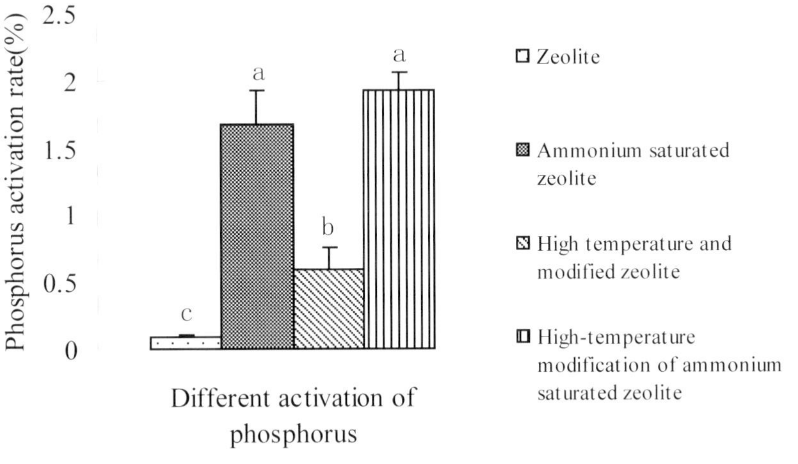

(a)

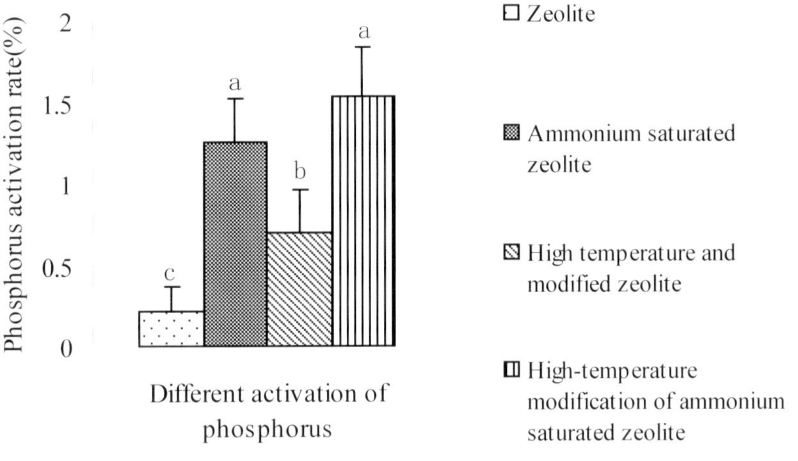

(b)

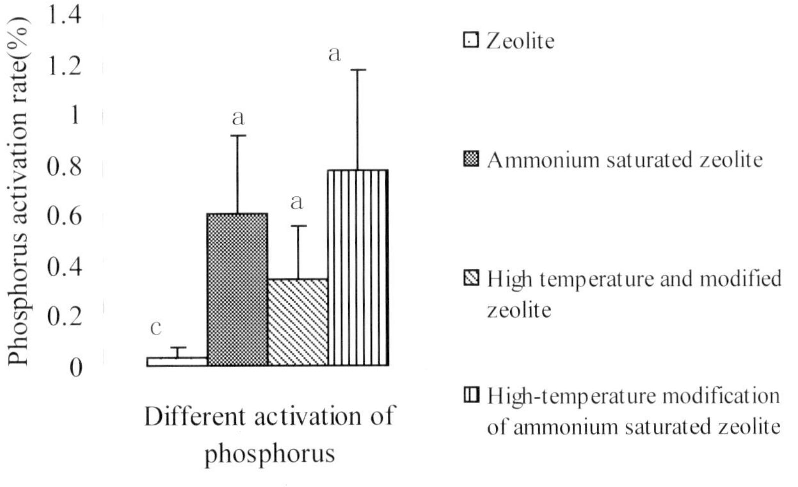

(c)

Figure 3. (a) The difference between activation rate of phosphorus in different modification treatment zeolite (PS high); (b) The difference between activation rate of phosphorus in different modification treatment zeolite (PS medium); (c) The difference between activation rate of phosphorus in different modification treatment zeolite (PS low).

mostly activate the fixed phosphorus in red soil. However, the phosphorus activation rate of high temperature modified ammonium saturated zeolite was quite different between three soil samples. The reasons are: firstly, total phosphorus content is different in these three soil samples. Secondly, available phosphorus content in the soil sample with high SP are the highest, which makes the free cation content that can fix available phosphorus relatively less, thus saturated NH_4^+ absorbed on high temperature modified zeolite could have cation exchange reaction with metal cation in insoluble phosphate, and the activation rate of phosphorus can be maximized.

4. Conclusions

The ability of zeolite in activating phosphorus of the red soil with different modified treatment was significantly higher than the ordinary unmodified one, and the treatment with high temperature modified and saturated with ammonium ion has the highest phosphorus activation amount. The absolute amount of activating phosphorus with different treatments has the order as follows: high-temperature modified ammonium saturated zeolite > ammonium saturated zeolite > high-temperature modified zeolite > ordinary zeolite.

For the tested soil samples with high, middle and low levels of rapid available phosphorus content, the increasing rate of soil phosphorus activation with different treatments has the same general trend. The increasing rate of modified phosphorus in high temperature ammonium saturated zeolite and ammonium saturated zeolite had a decline trend with time postponed while that for modified in high temperature zeolite and routine zeolite were rising gradually with time postponed. For ammonium saturated zeolite, the increasing rate of phosphorus in low available phosphorus content red soil was significantly faster than that in red soil with the higher content of available phosphorus. All these are related with their different activation mechanisms.

The four treatments by the usage in the soil with high, middle and low available phosphorus content have the same order of activation rate to phosphorus: high-temperature modified ammonium saturated zeolite > ammonium saturated zeolite > high temperature modified zeolite> ordinary zeolite, but the activation rate for the same modified zeolite showed difference in different phosphorus level soil. The order is: $SP_{high} > SP_{medium} > SP_{low}$.

Acknowledgements

Project supported by the science and technology innovation talent plan of Yunnan Province (No. 2015HC018); national water pollution prevention and treatment of major projects (No. 2012ZX07102-003).

References

[1] Schindler, D.W. (1997) The Evolution of Phosphorus Limitation in Lakes. *Science*, **195**, 260-262. https://doi.org/10.1126/science.195.4275.260

[2] He, Z.L., Yuan, K.N. and Zhu, Z.X. (1990) The Effect of Organic Anions on Phosphate Ad-

sorption. *Acta Pedologica Sinica*, **27**, 377-383. (In Chinese)

[3] Sakaih, T. (1991) Secretion of Acid Phosphatase by the Roots of Several Crop Species under Phosphorus-Deficient Conditions. *Soil Science and Plant Nutrition*, **37**, 129-140. https://doi.org/10.1080/00380768.1991.10415018

[4] Pomeioy, L.R., Simith, E.E. and Crant, C.M. (1965) The Exchange of Phosphorus between Estuarine Water and Sediments. *Limnology and Oceanography*, **10**, 167-172. https://doi.org/10.4319/lo.1965.10.2.0167

[5] Zhang, Y., Zheng, B.H., Liu, H.L., *et al.* (2005) The Traits Analysis of Nitrogen and Phosphorus Nutrients after Gorges Reservoir Impounding. *Water Resource Protection*, **21**, 23-26. (In Chinese)

[6] Xu, D.X., Qin, Y.W. and Zhang, L. (2009) Phosphorus Forms and Its Distribution Characteristics in Sediments and Soils of Water Level Fluctuating Zone of the Backwater Reach from Input River of Three Gorges Reservoir. *Environmental Science*, **5**, 83-90. (In Chinese)

[7] Barbarick, L., Lai, T.M. and Eberl, D.D. (1990) Exchange Fertilizer (Phosphate Rock Plus Ammonium-Zeolite) Effects on Sorghum-Sudangrass. *Soil Science Society American Journal*, **54**, 911-916. https://doi.org/10.2136/sssaj1990.03615995005400030050x

[8] Ando, H. and Mihara, C. (1996) The Fate of Ammonium Nitrogen Applied to Flooded Rice as Affected by Zeolite Addition. *Soil Science*, **42**, 531-538. https://doi.org/10.1080/00380768.1996.10416322

[9] Li, L.J. (1995) Ammonia Saturated Zeolite Increased the Yield of the Effect and the Influence of the Fertilizer Utilization Ratio. *Soils*, **3**, 147-150. (In Chinese)

[10] Li, L.J. and Zhang, L. (1996) Exchange Sex NH_4^+—Zeolite to Improve Soil p Force Research. *Soil Science Society of America Journal*, **27**, 139-140.

[11] Zorpas, A.A., Constantinides, T., Vlysside, A.G., *et al.* (2000) Heavy Metal Uptake by Natural Zeolite and Metals Partitioning in Sewage Sludge Compost. *Bioresource Technology*, **72**, 113-119. https://doi.org/10.1016/S0960-8524(99)00110-8

[12] Guan, L.Z. and Li, H.Z. (1992) The Effect of Increase of Natural Zeolite and Nitrogen and Phosphorus Nutrients and Some Fertility Regulation Mechanism of Nature. *Chinese Journal of Soil Science*, **23**, 205-208. (In Chinese)

[13] Li, L.J. and Yin, X. (1999) Ammonium Saturated Zeolite-Apatite Soil Phosphorus Release the Role and Application. *International Chinese Agriculture Science and Technology Annual Meeting*.

[14] Bogumil, K. and Barbara, G. (2010) Mechanochemical Activation of Phosphate Rock in Presence of Zeolites. *Praemysl Chemiczny*, **89**, 414-418.

[15] Abd-Alla, M.H. (1994) Use of Organic Phosphorus by Rhizobium Leguminosarum Biovar Viceae Phosphatases. *Biology and Fertility of Soils*, **18**, 216-218. https://doi.org/10.1007/BF00647669

5

Air Pollution in Major Chinese Cities: Some Progress, But Much More to Do

Dorrit H. Lowsen[1], George A. Conway[2]

[1]U.S. Department of State, Beijing, China

[2]Division of Global Health Protection, Center for Global Health, US Centers for Disease Control and Prevention, Atlanta, GA, USA

Email: dorritg@gmail.com, george.conway@deschutes.org

Abstract

Background: Ambient (outdoor) air pollution has been implicated as a major cause of acute cardiovascular and pulmonary illnesses and increased risk for acute and chronic effects after chronic exposures, including mortality and morbidity. In 2008, due to persistent health concerns about its workforce and their dependents, the US Mission in China began monitoring air quality at the US Embassy in Beijing. Subsequently, monitoring stations were also established at US consulates at Shanghai (2011), Guangzhou (2011), Chengdu (2012), and Shenyang (2013). **Objectives:** To determine whether there have been definable trends in air quality in these five Chinese cities. **Methods:** Air monitoring results from each locale for accumulated PM2.5 particulate matter were calculated hourly. Accumulated data were organized, culled using a standardized set of heuristics, and analyzed for trends. **Results:** China's capital city, Beijing, experienced decreased PM2.5 from 2013 through 2015, but no significant long-term downward trend from 2008 through 2015. Shanghai has not shown any definable air quality trend since 2012. Chengdu experienced some improvement in air quality since 2013, but none discernible from 2012 through 2015. Guangzhou had generally better air quality, and a downward trend since 2012. Shenyang experienced increasingly severe air pollution from 2013 through 2015. **Conclusion:** There appear to have been recent tangible, though modest, improvements in air quality in three large Chinese cities: Beijing, Chengdu, and Guangzhou, but no apparent progress in Shanghai, and a worrisome decline in air quality observed in Shenyang. Despite recent progress, there is a long way to go before even the cities which show improvement reach Chinese standards.

Keywords

Environmental Monitoring, Air Pollution/Air Q, Ambient Air Quality, China, PM2.5

1. Introduction

In 2008, due to persistent health concerns about its workforce and dependents, the US Mission in China began monitoring air quality at the US Embassy in Beijing. Subsequently, monitoring stations were also established at US consulates in Shanghai (2011), Guangzhou (2011), Chengdu (2012), and Shenyang (2013). Hourly values are available online (and with a mobile app) approximately fifteen minutes after each hour of the clock. Interest in this program increased during a series of acute, severe air pollution events which occurred during persistent "thermal" inversion events (multi-day periods with trapping of air in layers) in northeastern China in 2013, the largest scale affecting many provinces and lasting from approximately January 10 through 23 (with some brief lulls), which were widely covered by the domestic and international press, and given the moniker of "airpocalypse" by many expatriates [1] [2]. China's discussions and efforts to curtail air pollution increased markedly after this signal event, with expansion of domestic air monitoring, creation of air pollution alert thresholds and communications systems, and more widespread posting of real-time air quality observations publicly available on websites for more Chinese cities [3] [4] [5]. Sales of filtering face masks and household air cleaners increased markedly in many Chinese cities. Severe air pollution events also occurred in the Harbin city area in October of 2013 [6] [7] and Shanghai in November, 2013 [8]. A prior publication [9] analyzing diplomatic facility monitoring data from China reported marked diurnal and seasonal variations in air pollution.

2. Background

Ambient (outdoor) air pollution has been implicated as a major cause of acute cardiovascular and pulmonary illnesses for many years during acute air inversion events, most notoriously in the United States in Donora, PA, 1948 [10] and in London, UK in 1952 [11]. Since then evidence has accumulated for an increased risk for acute and chronic effects after chronic exposures, including mortality [12] [13] [14] and morbidity [15] [16] [17]. PM-related mortality and morbidity studies have been conducted in some severely air-polluted cities in China [18] [19]. Epidemiologic analyses of the effects of PM-related to disease morbidity have been widely conducted in hospitals, special populations and people with medical conditions [20] [21] [22].

Most worrisome among published work may be the findings of a long-term cohort study of children exposed to air pollutants in southern California, which implicates increased exposure to such pollutants in delayed lung development, an increase of new cases and exacerbations of asthma and bronchitis, and airway inflammation [23]. Estimates of negative subclinical respiratory effects, such as decreased lung function detected by spirometry, and symptoms of PM air pollution have been made in other nations [24] [25] [26] [27], as in China [28] [29] [30].

A substantial [25] [28] [29] [31] [32] [33] [34] literature describes biomedical and imputed long-term effects of air pollution on the human central nervous system from cellular to systemic effects, but with uncertainties remaining about the mechanism for such effects [32].

Over the last decade, studies have been published examining how acutely the onset of symptoms and/or deleterious physiologic/pathologic changes may occur after exposure to air pollutants. For example, a double-blinded, randomized crossover study found that "Brief exposure to dilute diesel exhaust promotes myocardial ischemia and inhibits endogenous fibrinolytic capacity in men with stable coronary disease... within 15 minutes from the start of exposure to dilute diesel exhaust." [35]. While this study alone (and unreplicated) might not fully implicate such exposure, it is consistent with other published evidence for exposure to air pollution worsening symptoms of angina pectoris in a multi-center study [36] and exacerbating myocardial ischemia during exercise [37] [38].

Fine, or PM2.5 particles, markedly smaller than PM 10, can travel into the end airways and alveoli, and many of the components may directly enter the bloodstream from there. Thus, the effects of PM 2.5 may be more insidious than larger particles, with marked systemic effects. Recent studies demonstrate rapid electrocardiographic, physiologic and symptomatic changes after exposure to PM2.5 particles and related aerosols such as diesel fumes. "These effects do not appear to be present immediately following the exposure but can be demonstrated from as early as 2 h after the exposure...." [39]. The well-documented relationship between PM2.5 and health effects, as well as the synthesis of many hazardous air quality fractions in this single measure, is why many air quality monitoring systems collect real-time PM2.5 as a sole or key indicator of air quality, and why we focus this study on that [aggregate] measure [40].

Thus, sudden changes in the severity of air pollution may have rapid effects on persons with asthma and appear to have very specific, quantifiable effects on persons with existing coronary disease, particularly those with ischemia and a history of angina pectoris. This, in turn, would support the need for as frequent and accurate measurements of air quality as possible, and the sharing of such information on a real-time basis. This information may be useful in providing warnings to susceptible individuals when conditions deteriorate, and reassurance that it is now safer to go outside to exercise when those conditions improve.

3. Objective

To determine whether there have been definable trends in air quality in these five Chinese cities.

4. Methods

Air monitoring is conducted around the clock in each locale using one or two Met One Instruments Beta Attenuation monitors (BAM 1020). Monitor placement was determined in consultation with experts from the US Environmental Protection Agency (EPA) and varies by location, but is generally consistent with air mass placement with good exposure from all sides as defined in Table 7-1 of the EPA's Quality Assurance Handbook for Air Pollution Measurement Systems [41]. Monitors are serviced and calibrated monthly with additional servicing provided as needed in the event of unexplained downtime or anomalous readings. The monitors take hourly measurement of accumulated PM2.5 particulate matter (which can include carbon [in elemental and organic forms], sulfate [SO_2] and nitrate [NO_x] [often dissolved in water droplets, particularly during foggy and static air conditions], ammonium, silicon, and sodium ion, as well crustal matter and trace elements, including heavy metals) [42] [43] [44] used to calculate an average concentration in μg/m^3 for the hour, consistent with EPA standards for measurement schedule [41]. These measurements are converted using a standardized algorithm to a stepwise-linear scale, the Air Quality Index (AQI) using the formula $AQI = [(AQI_H - AQI_L)/(BP_H - BP_L)](Con - BP_L) + AQI_L$ where Con represents the measured concentration, BP_H and BP_L represent the top and bottom values for the breakpoint range from **Table 1** in which Con falls, and AQI_H and AQI_L represent the top and bottom of the corresponding AQI range [45]. The resulting AQIs, along with the (underlying) PM2.5 concentrations are posted to US Mission China website (URL: http://stateair.net) and Twitter feeds (Handles: @BeijingAir, @CGShanghaiAir, @Guangzhou_air, @CGChengduAir) along with health/activity warnings as indicated.

Table 1. Air quality index breakpoints [45].

Category	Concentration Breakpoint (BP)		Air Quality Index (AQI)	
	Low	High	Low	High
Good	0	12	0	50
Moderate	12.1	35.4	51	100
Unhealthy for Sensitive Groups	35.5	55.4	101	150
Unhealthy	55.5	150.4	151	200
Very Unhealthy	150.5	250.4	201	300
Hazardous	250.5	500.4	301	500

The raw data presented here are all available for download at the US Mission China air quality website, http://stateair.net.

The data were organized using a simple Microsoft Excel spreadsheet (various versions were used over the course of the work) with Excel's built in functionality used to provide the following clean up and statistical analysis.

First, data were cleaned to close gaps. Single hour gaps in the raw data were filled using the arithmetic mean of the two surrounding data points. Multi-hour gaps in the raw data were not filled. The hourly data were then used to calculate daily, monthly, quarterly, half yearly, and annual averages (arithmetic mean), excluding from analysis any periods of time with too many missing data points resulting from multi-hour gaps. Time periods were excluded if data were available for fewer than half of the expected single-hour readings. The expected number of single-hour readings were calculated as 24 * approximate number of days in the period (year = 365, half year = 182, quarter = 91, month = 30, day = 1). Finally, trends were calculated across comparable time periods from year to year (e.g. first quarter of each year in the data set) using the linear regression tool from Excel's Data Analysis toolkit which utilizes LINEST function to fit a line using the least squares method [46] [47].

In addition to examining trends, various intervals were evaluated for attainment of Chinese national air quality standards [48], as well as how often each city would meet international guidelines set by the World Health Organization [49]. Given the purpose of the US Mission China's monitoring program, the percentage of days which had mean values falling below US air quality standards [50] were also calculated.

5. Results

China's capital city, Beijing, experienced a decrease in PM2.5 from 102 μg/m^3 in 2013 to 83 μg/m^3 in 2015, the lowest annual average since U.S. Embassy monitoring began in 2008. However, despite these apparent improvements in recent years, the data do not show a statistically significant downward trend since the beginning of the monitoring program (**Figure 1** and **Table 2**). For 2015, 28.5 % of Beijing's days met Chinese air quality standards, 20.1 % of days met WHO guidelines, and 6.9 % of days met US air quality standards (**Figure 2** and **Table 3**).

China's most populous city, Shanghai, while it does not usually suffer as severe air pollution on average as Beijing, has not shown any clear trend since monitoring began in 2012 (**Figure 1** and **Table 2**), nor improvement in attainment of Chinese air quality standards over that interval (**Figure 2** and **Table 3**).

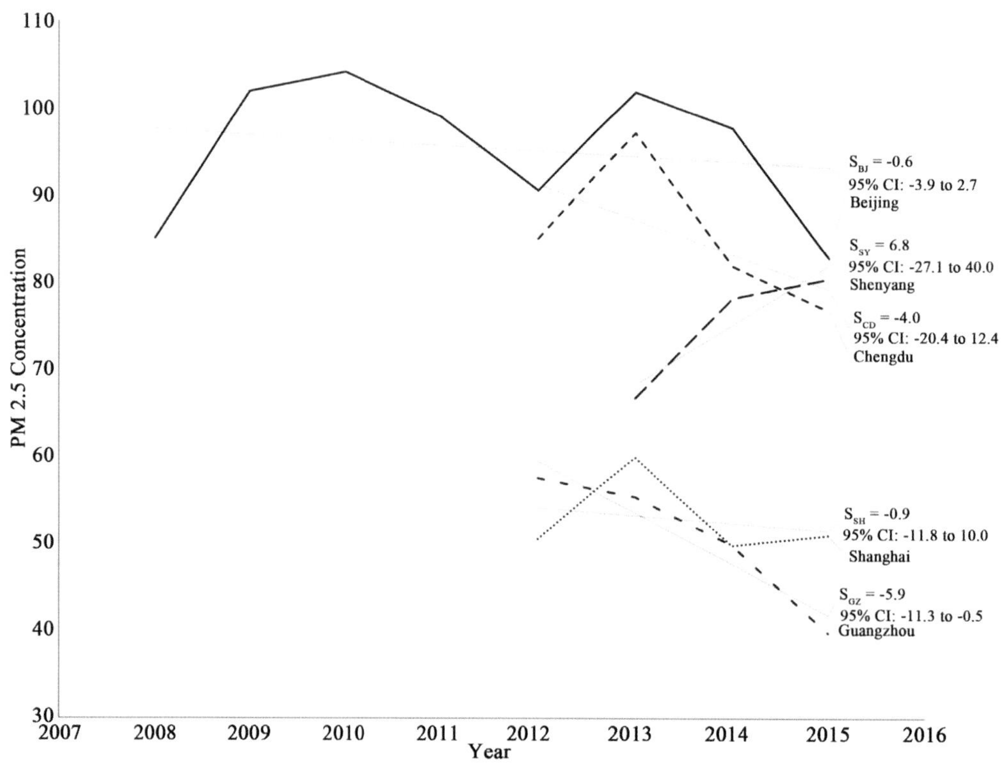

Figure 1. Annual average PM2.5 concentrations, 2008-2015.

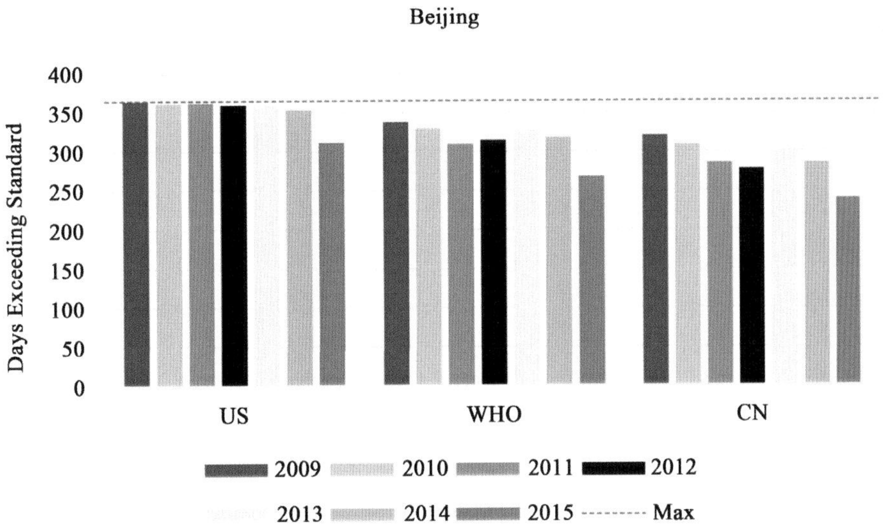
Beijing
Days Exceeding Standard
400
350
300
250
200
150
100
50
0
US
WHO
CN
2009
2010
2011
2012
2013
2014
2015
Max

(a)

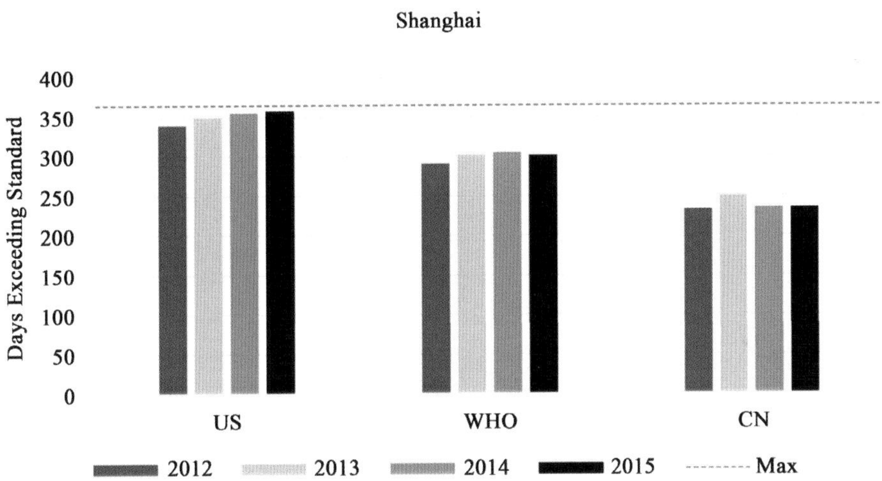
Shanghai
Days Exceeding Standard
400
350
300
250
200
150
100
50
0
US
WHO
CN
2012
2013
2014
2015
Max

(b)

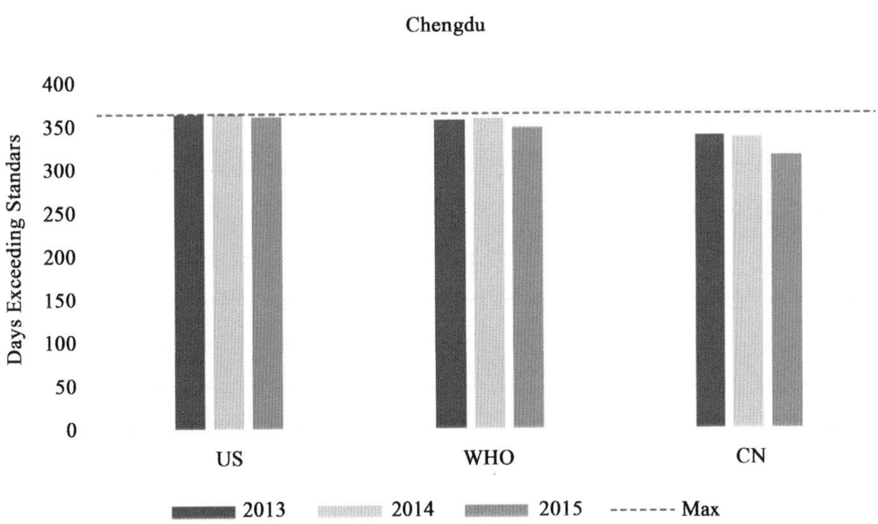
Chengdu
Days Exceeding Standars
400
350
300
250
200
150
100
50
0
US
WHO
CN
2013
2014
2015
Max

(c)

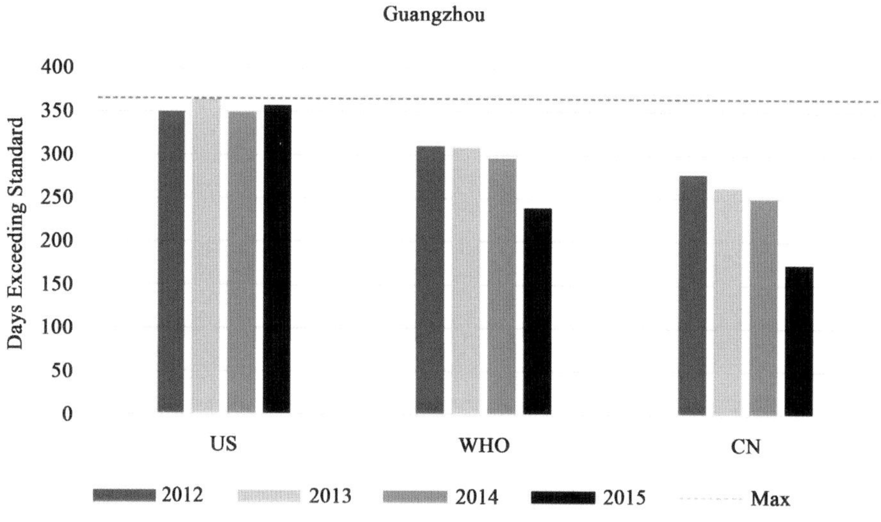

(d)

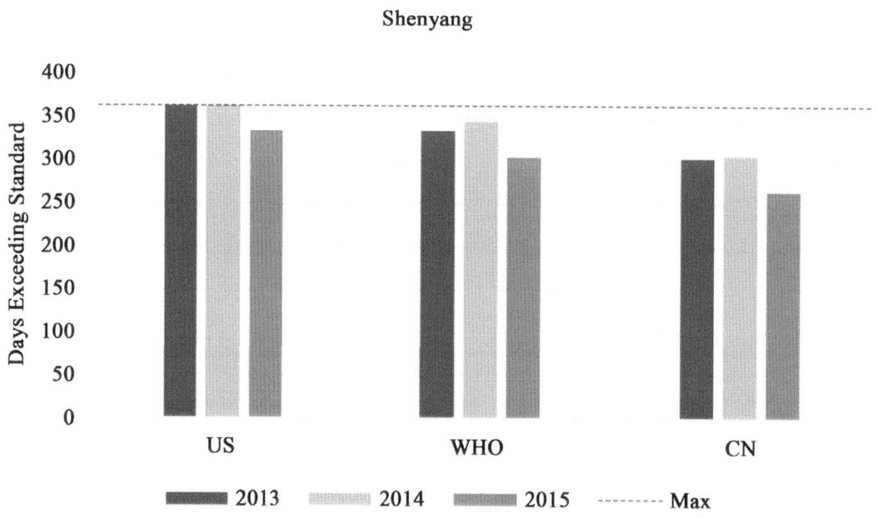

(e)

Figure 2. Number of days each year exceeding current US and Chinese air quality standards and WHO guidelines.

Table 2. Annual average PM2.5 concentrations 2008-2015, Chinese cities with US Mission monitoring.

Year	Beijing	Shanghai	Chengdu	Guangzhou	Shenyang
2008	85				
2009	102				
2010	104				
2011	99				
2012	91	51	85	58	
2013	102	60	97	55	67
2014	98	50	82	50	78
2015	83	51	77	40	80

Table 3. Percentage of days each year meeting current US and Chinese air quality standards and WHO guidelines.

China	2008	2009	2010	2011	2012	2013	2014	2015	Mean
Beijing	18%	15%	17%	23%	25%	17%	22%	29%	21%
Shanghai					37%	32%	36%	37%	35%
Chengdu					6%	8%	8%	13%	9%
Guangzhou				14%	32%	29%	33%	53%	37%
Shenyang						27%	16%	22%	21%

WHO	2008	2009	2010	2011	2012	2013	2014	2015	Mean
Beijing	9%	10%	10%	16%	15%	10%	13%	20%	13%
Shanghai					21%	17%	17%	18%	18%
Chengdu					2%	2%	1%	4%	3%
Guangzhou				6%	20%	16%	20%	35%	22%
Shenyang						13%	5%	10%	9%

US	2008	2009	2010	2011	2012	2013	2014	2015	Mean
Beijing	2%	0%	1%	1%	2%	1%	3%	7%	2%
Shanghai					7%	4%	2%	2%	4%
Chengdu					2%	0%	0%	1%	0%
Guangzhou				0%	6%	0%	5%	2%	3%
Shenyang						0%	0%	1%	0%

Chengdu, the westernmost large city in China and capital of Szechuan Province, has experienced an apparent improvement in air quality since 2013, but not since 2012 (**Figure 1** and **Table 2**). The number of days meeting air quality standards remains low (**Figure 2** and **Table 3**).

Guangzhou, the southernmost city of China with a US consulate, has generally better air quality than the other cities monitored, with a downward trend since 2012 (**Figure 1** and **Table 2**), but still exceeded Chinese air quality standards 47% of days in 2015, and seldom fell below US standards (**Figure 2** and **Table 3**).

Shenyang, in northeastern China, while generally having somewhat better air quality throughout this interval than much-larger Beijing, experienced increasingly severe air pollution from 2013 through 2015, with its mean PM2.5 concentrations converging on those of Beijing (**Figure 1** and **Table 2**) by 2015.

6. Discussion

A number of factors may affect the conclusions drawn from the data reported in this study. First, each US Mission China facility measures air quality in its immediate vicinity, so may not be fully representative of its city as a whole. In fact, in November of 2015 the Beijing Environmental Protection Bureau announced that Beijing's air quality in the first ten months of 2015 had improved 21.8% over the same period in 2014 [51], while the U.S. Embassy's data showed a more dramatic improvement of 28.9%.

It is outside the scope of this study to evaluate causes of improvement (or decline) in air pollution although it is worth noting for future research some factors which should be considered. It is generally accepted that meteorological conditions including wind, temperature, and humidity can affect local air pollution and its measurement [4] [28]. With the right data and analysis, it might be possible to evaluate whether the recent improvement trend observed, particularly in Beijing in 2014 in preparation for the Asia-Pacific Economic Conference and 2015 before the "70th Anniversary of the Victory of the Chinese People's War of Resistance Against Japanese Aggression and the World Anti-Fascist War" [52], directly reflects improvements resulting from successful limitation of pollution inputs (e.g., coal combustion for power and heat generation reductions and/or smokestack pollution controls, diesel exhaust from heavy transport vehicles, automotive exhausts, construction dusts, etc.) or from other factors, such as air movement (wind) and the pace of industrial production and consumption due to economic shifts. Such analysis would help us to more effectively discern whether recent rates of improvement are likely to be sustainable or are primarily the result of either factors outside human control (e.g. weather) or one-time changes (e.g. Beijing's recent closure of most coal-fired power plants within the city's core) which represent real gains, but not on-going year-over-year improvements. If recent rates of improvement in Beijing, Chengdu, and Guangzhou prove to be sustainable, those cities could reach Chinese, and even international air quality standards [48] within just a few years. If, on the other hand, recent improvements prove to have resulted from weather conditions, or as these authors think possible, from specific but not repeatable control measures such as closure of coal-fired power plants, or ones which may strain available public acceptance and transportation infrastructure, e.g., implementation of alternate-day driving restrictions, then the rate of improvement would likely level off, leaving attainment of air quality standards unpredictably far in the future.

It should also be noted that PM2.5 measurements may not capture the full hazards for air pollution, which are also posed by ozone, carbon monoxide, volatile organic compounds, ultrafine particles and possibly as-yet-unknown fractions.

A fractionated analysis of the PM2.5 measurements described here would provide a more granular picture, and might more precisely reflect relative hazards and trends, but are beyond the scope of this study. Interestingly, a published study of such fractions in Beijing during the highly polluted winter of 2012-2013 demonstrated rapid conversion of gas phase to particle phase for nitrates and (especially) sulfates [42].

Most importantly, this study only measured the severity of air pollution through time. The arduous process of gathering a wide range of health data and examining their frequency and temporal relationship to worsening and/or improvements in air pollution should continue apace.

7. Conclusion

These data provide longitudinal detail for air quality in multiple Chinese cities. There do appear to have been tangible, though modest, recent improvements in air quality in three large Chinese cities, Beijing, Chengdu, and Guangzhou, but no apparent progress in Shanghai, and a worrisome decline in air quality observed in Shenyang. Despite recent progress, there is a long way to go before even the cities which show improvement reach Chinese standards (let alone the more rigorous goals set by the WHO guidelines and US standards). If recently-observed rates of improvement persist, that could happen quickly, but our data are insufficient to conclude that those rates are durable. Relying on the longer history from Beijing, which shows little or no overall improvement since 2008, despite recent gains over peak pollution levels, the data suggest it could be a very long time before China reaches safe air quality. This could be interpreted as putting many adults at increased risk of cardiopulmonary and other diseases, and consigning a generation of Chinese children in those locales to diminished pulmonary capacities and other hazards.

Acknowledgements

The findings and conclusions in this paper are those of the author(s) and do not necessarily represent the views of the Centers for Disease Control and Prevention.

John K. Whittlesey, Deputy Counselor, Environment, Science, Technology & Health, US Embassy, Beijing; Matthew Paschke, US Department of State, Baghdad, Iraq and Erica Thomas, former Environment, Science, Technology & Health Chief, US Embassy Beijing; former Deputy Chiefs of Mission Daniel J. Kritenbrink and Robert S. Wang, US Embassy, Beijing; Ambassador Max Baucus and former Ambassador Gary Locke, US Embassy Beijing. Robert E Fontaine, MD, MSc, Senior Adviser for the Field Epidemiology Training Program Branch, CDC; Ronald L. Moolenaar, MD, MPH, CAPT, USPHS, Country Director, China Programs, Center for Global Health, CDC; Dr. Paul Garbe, Air Pollution Branch, National Center for Environmental Health, CDC, Atlanta, GA.

References

[1] Lim, L. (2013) Beijing's "Airpocalypse" Spurs Pollution Controls, Public Pressure. Morning Edition, National Public Radio. http://www.npr.org/2013/01/14/169305324/beijings-air-quality-reaches-hazardous-levels

[2] Wu, X. (2013) More Serious Air Pollution in China Cities and the Haze Cover in an Area of 130 km^2. China New Network.

[3] China National Environmental Monitoring Centre (2013) Air Quality Index (AQI) of 74 Cities in China, January 2013. China National Environmental Monitoring Centre.

[4] Ma, Z., Hu, X., Sayer, A.M., Levy, R., Zhang, Q., Xue, Y., *et al.* (2015) Satellite-Based Spatiotemporal Trends in PM2.5 Concentrations: China, 2004-2013. *Environmental Health Perspectives*, **124**, 184-192. https://doi.org/10.1289/ehp.1409481

[5] Stoerk, T. (2016) Statistical Corruption in Beijing's Air Quality Data Has Likely Ended in 2012. *Atmospheric Environment*, **127**, 365-371. https://doi.org/10.1016/j.atmosenv.2015.12.055

[6] Nunez, C. (2013) Harbin Smog Crisis Highlights China's Coal Problem. National Geographic.

[7] Wong, E. (2013) Response to a City's Smog Points to a Change in Chinese Attitude. The New York Times.

[8] Saul, H. (2013) Shanghai Smog: Air Pollution Covers City. The Independent.

[9] San Martini, F.M., Hasenkopf, C.A. and Roberts, D.C. (2015) Statistical Analysis of PM2.5 Observations from Diplomatic Facilities in China. *Atmospheric Environment*, **110**, 174-185. https://doi.org/10.1016/j.atmosenv.2015.03.060

[10] Guenther, E. (2011) Donora Death Fog: The Crisis That Led to Modern Air Pollution Laws. http://www.aiche.org/chenected/2011/10/donora-death-fog-crisis-led-modern-air-pollution-laws

[11] Bell, M.L. and Davis, D.L. (2001) Reassessment of the Lethal London Fog of 1952: Novel Indicators of Acute and Chronic Consequences of Acute Exposure to Air Pollution. *Environmental Health Perspectives*, **109**, 389-394.

[12] Katsouyanni, K., Touloumi, G., Spix, C., Schwartz, J., Balducci, F., Medina, S., *et al.* (1997) Short-Term Effects of Ambient Sulphur Dioxide and Particulate Matter on Mortality in 12 European Cities: Results from Time Series Data from the APHEA Project. *BMJ*, **314**, 1658-1663. https://doi.org/10.1136/bmj.314.7095.1658

[13] Krall, J.R., Anderson, G.B., Dominici, F., Bell, M.L. and Peng, R.D. (2013) Short-Term Exposure to Particulate Matter Constituents and Mortality in a National Study of U.S. Urban Communities. *Environmental Health Perspectives*, **121**, 1148-1153. https://doi.org/10.1289/ehp.1206185

[14] Weichenthal, S., Villeneuve, P.J., Burnett, R.T., van Donkelaar, A., Martin, R.V., Jones, R.R., *et al.* (2014) Long-Term Exposure to Fine Particulate Matter: Association with Non-accidental and Cardiovascular Mortality in the Agricultural Health Study Cohort. *Environmental Health Perspectives*, **122**, 609-615. https://doi.org/10.1289/ehp.1307277

[15] Jakubiak-Lasocka, J., Lasocki, J. and Badyda, A.J. (2014) The Influence of Particulate Matter on Respiratory Morbidity and Mortality in Children and Infants. In: Pokorski, M., Ed., *Environmental Biomedicine, Advances in Experimental Medicine and Biology Vol.* 849, Springer International Publishing, Switzerland, 39-48. https://doi.org/10.1007/5584_2014_93

[16] Lippmann, M., Ito, K., Nádas, A. and Burnett, R.T. (2000) Association of Particulate Matter Components with Daily Mortality and Morbidity in Urban Populations. Research Reports (Health Effects Institute), No. 95, 5-72.

[17] Wolf, K., Stafoggia, M., Cesaroni, G., Andersen, Z.J., Beelen, R., Galassi, C., *et al.* (2015) Long-Term Exposure to Particulate Matter Constituents and the Incidence of Coronary Events in 11 European Cohorts. *Epidemiology*, **26**, 565-574. https://doi.org/10.1097/EDE.0000000000000300

[18] Wong, C.-M., Vichit-Vadakan, N., Kan, H. and Qian, Z. (2008) Public Health and Air Pollution in Asia (PAPA): A Multicity Study of Short-Term Effects of Air Pollution on Mortality. *Environmental Health Perspectives*, **116**, 1195-1202. https://doi.org/10.1289/ehp.11257

[19] Zhang, L., Chen, X., Xue, X., Sun, M., Han, B., Li, C., *et al.* (2014) Long-Term Exposure to High Particulate Matter Pollution and Cardiovascular Mortality: A 12-Year Cohort Study in Four Cities in Northern China. *Environment International*, **62**, 41-47. https://doi.org/10.1016/j.envint.2013.09.012

[20] Linares, C. and Díaz, J. (2009) Impact of Particulate Matter with Diameter of Less than 2.5 Microns [PM2.5] on Daily Hospital Admissions in 0-10-Year-Olds in Madrid, Spain [2003-

2005]. *Gaceta Sanitaria*, **23**, 192-197. https://doi.org/10.1016/j.gaceta.2008.04.006

[21] Lin, C.-M. and Kuo, H.-W. (2013) Sex-Age Differences in Association with Particulate Matter and Emergency Admissions for Cardiovascular Diseases: A Hospital-Based Study in Taiwan. *Public Health*, **127**, 828-833. https://doi.org/10.1016/j.puhe.2013.04.010

[22] Zhu, R., Chen, Y., Wu, S., Deng, F., Liu, Y. and Yao, W. (2013) The Relationship between Particulate Matter (PM_{10}) and Hospitalizations and Mortality of Chronic Obstructive Pulmonary Disease: A Meta-Analysis. *COPD: Journal of Chronic Obstructive Pulmonary Disease*, **10**, 307-315. https://doi.org/10.3109/15412555.2012.744962

[23] Chen, Z., Salam, M.T., Eckel, S.P., Breton, C.V. and Gilliland, F.D. (2015) Chronic Effects of Air Pollution on Respiratory Health in Southern California Children: Findings from the Southern California Children's Health Study. *Journal of Thoracic Disease*, **7**, 46-58.

[24] Álvarez Miño, L. and Salazar Ceballos, A. (2013) Respiratory Symptoms and Lung Function in Children Aged 6 - 14 Years and Their Relationship with Particulate Matter PM_{10} in Santa Marta, Colombia. *Revista Española de Salud Pública*, **87**, 239-246.

[25] Block, M.L., Elder, A., Auten, R.L., Bilbo, S.D., Chen, H., Chen, J.-C., *et al.* (2012) The Outdoor Air Pollution and Brain Health Workshop. *Neurotoxicology*, **33**, 972-984.

[26] Hrubá, F., Fabiánova, E., Koppová, K. and Vandenberg, J. (2001) Childhood Respiratory Symptoms, Hospital Admissions, and Long-Term Exposure to Airborne Particulate Matter. *Journal of Exposure Analysis and Environmental Epidemiology*, **11**, 33-40. https://doi.org/10.1038/sj.jea.7500141

[27] Karakatsani, A., Analitis, A., Perifanou, D., Ayres, J.G., Harrison, R.M., Kotronarou, A., *et al.* (2012) Particulate Matter Air Pollution and Respiratory Symptoms in Individuals Having Either Asthma or Chronic Obstructive Pulmonary Disease: A European Multicentre Panel Study. *Environmental Health*, **11**, 75. https://doi.org/10.1186/1476-069X-11-75

[28] Cui, L., Li, X., Zhang, Y., *et al.* (2014) Study on the Health Effect of Respiratory Acute and Severe Symptoms during a Haze Event in January, 2013, Jinan City. *Disease Surveillance*, **29**, 852-856.

[29] Cui, L., Zhou, J., Zhang, J., *et al.* (2014) Analysis on the Common and Severe Acute Symptoms Health Effect during Haze Event Period. *Journal of Environmental Health*, **31**, 600-603.

[30] Pan, G., Zhang, S., Feng, Y., Takahashi, K., Kagawa, J., Yu, L., *et al.* (2010) Air Pollution and Children's Respiratory Symptoms in Six Cities of Northern China. *Respiratory Medicine*, **104**, 1903-1911. https://doi.org/10.1016/j.rmed.2010.07.018

[31] Calderón-Garcidueñas, L., Solt, A.C., Henriquez-Roldan, C., Torres-Jardon, R., Nuse, B., Herritt, L., *et al.* (2008) Long-Term Air Pollution Exposure Is Associated with Neuroinflammation, an Altered Innate Immune Response, Disruption of the Blood-Brain Barrier, Ultrafine Particulate Deposition, and Accumulation of Amyloid β-42 and α-Synuclein in Children and Young Adults. *Toxicologic Pathology*, **36**, 289-310. https://doi.org/10.1177/0192623307313011

[32] Calderón-Garcidueñas, L., Reed, W., Maronpot, R.R., Henríquez-Roldán, C., Delgado-Chavez, R., Calderón-Garcidueñas, A., *et al.* (2004) Brain Inflammation and Alzheimer's-Like Pathology in Individuals Exposed to Severe Air Pollution. *Toxicologic Pathology*, **32**, 650-658. https://doi.org/10.1080/01926230490520232

[33] Levesque, S., Surace, M.J., McDonald, J. and Block, M.L. (2011) Air Pollution & the Brain: Subchronic Diesel Exhaust Exposure Causes Neuroinflammation and Elevates Early Markers of Neurodegenerative Disease. *Journal of Neuroinflammation*, **8**, 105. https://doi.org/10.1186/1742-2094-8-105

[34] Segalowitz, S.J. (2008) Public Health, Brain Health, and the Dangers of Air Pollution for Neural Development. *Brain and Cognition*, **68**, 115-116. https://doi.org/10.1016/j.bandc.2008.08.011

[35] Mills, N.L., Törnqvist, H., Gonzalez, M.C., Vink, E., Robinson, S.D., Söderberg, S., *et al.* (2007) Ischemic and Thrombotic Effects of Dilute Diesel-Exhaust Inhalation in Men with Coronary Heart Disease. *The New England Journal of Medicine*, **357**, 1075-1082. https://doi.org/10.1056/NEJMoa066314

[36] von Klot, S., Peters, A., Aalto, P., Bellander, T., Berglind, N., D'Ippoliti, D., *et al.* (2005) Ambient Air Pollution Is Associated with Increased Risk of Hospital Cardiac Readmissions of Myocardial Infarction Survivors in Five European Cities. *Circulation*, **112**, 3073-3079. https://doi.org/10.1161/CIRCULATIONAHA.105.548743

[37] Gold, D.R., Litonjua, A.A., Zanobetti, A., Coull, B.A., Schwartz, J., MacCallum, G., *et al.* (2005) Air Pollution and ST-Segment Depression in Elderly Subjects. *Environmental Health Perspectives*, **113**, 883-887. https://doi.org/10.1289/ehp.7737

[38] Pekkanen, J., Peters, A., Hoek, G., Tiittanen, P., Brunekreef, B., de Hartog, J., *et al.* (2002) Particulate Air Pollution and Risk of ST-Segment Depression during Repeated Submaximal Exercise Tests among Subjects with Coronary Heart Disease: The Exposure and Risk Assessment for Fine and Ultrafine Particles in Ambient Air (ULTRA) Study. *Circulation*, **106**, 933-938. https://doi.org/10.1161/01.CIR.0000027561.41736.3C

[39] Miller, M.R., Shaw, C.A. and Langrish, J.P. (2012) From Particles to Patients: Oxidative Stress and the Cardiovascular Effects of Air Pollution. *Future Cardiology*, **8**, 577-602. https://doi.org/10.2217/fca.12.43

[40] Nuvolone, D., Balzi, D., Chini, M., Scala, D., Giovannini, F. and Barchielli, A. (2011) Short-Term Association between Ambient Air Pollution and Risk of Hospitalization for Acute Myocardial Infarction: Results of the Cardiovascular Risk and Air Pollution in Tuscany (RISCAT) Study. *American Journal of Epidemiology*, **174**, 63-71. https://doi.org/10.1093/aje/kwr046

[41] U.S. Environmental Protection Agency, Office of Air Quality Planning and Standards, Air Quality Assessment Division (2013) QA Handbook for Air Pollution Measurement Systems, Volume II: Ambient Air Quality Monitoring Program.

[42] Quan, J., Tie, X., Zhang, Q., Liu, Q., Li, X., Gao, Y., *et al.* (2014) Characteristics of Heavy Aerosol Pollution during the 2012-2013 Winter in Beijing, China. *Atmospheric Environment*, **88**, 83-89. https://doi.org/10.1016/j.atmosenv.2014.01.058

[43] Bell, M.L., Ebisu, K. and Peng, R.D. (2011) Community-Level Spatial Heterogeneity of Chemical Constituent Levels of Fine Particulates and Implications for Epidemiological Research. *Journal of Exposure Science and Environmental Epidemiology*, **21**, 372-384. https://doi.org/10.1038/jes.2010.24

[44] Martins, V., Moreno, T., Minguillón, M.C., van Drooge, B.L., Reche, C., Amato, F., *et al.* (2016) Origin of Inorganic and Organic Components of $PM_{2.5}$ in Subway Stations of Barcelona, Spain. *Environmental Pollution*, **208**, 125-136. https://doi.org/10.1016/j.envpol.2015.07.004

[45] Mintz, D. (2013) Technical Assistance Document for the Reporting of Daily Air Quality—The Air Quality Index (AQI). United States Environmental Protection Agency.

[46] Microsoft (2013) LINEST Function. Microsoft Office Support.

[47] Microsoft (2013) Use the Analysis ToolPak to Perform Complex Data Analysis. Microsoft Office Support.

[48] Ministry of Environmental Protection of the People's Republic of China (2012) Ambient

Air Quality Standards, 2012.

[49] World Health Organization (2006) WHO Air Quality Guidelines for Particulate Matter, Ozone, Nitrogen Dioxide and Sulfur Dioxide.

[50] United States Environmental Protection Agency (2012) Revised Air Quality Standards for Particle Pollution and Updates to the Air Quality Index (AQI).

[51] People's Daily (2015) Beijing PM2.5 Drops 21.8% in the First 10 Months.

[52] China Daily (2015) Beijing's Efforts to Control Air Pollution Start to Pay off.

6

Laboratory Evaluation of the Impact of Contaminants on Soil Resistivity and the Consequent Effect on Plant's Growth

Wilfred N. Igboama[1], Nicholas U. Ugwu[2*]

[1]Department of Physics, Federal University, Oye-Ekiti, Nigeria
[2]Department of Physics, University of Botswana, Gaborone, Botswana
Email: *nuumay665@gmail.com

Abstract

This study is focused on investigation of the resistivity property of polluted soils and the consequent effect on the growth rate of plants. The method involves construction of an inexpensive apparatus for measuring the electrical resistivity of polluted soil in the laboratory based on application of Ohm's law. Some pollutants such as petrol, kerosene and brine were added to the soil samples and the electrical resistivity was determined. The results showed that the quantity of petrol and kerosene added to the soil were directly proportional to the resistivity of the soil while the salt concentration varied inversely with resistivity. On the other hand, the study showed that Kerosene, Petrol and Salt solution of different concentrations have an adverse effect on the growth and development of the bean plants.

Keywords

Ohm's Law, Pollutants, Resistivity, Kerosene, Petrol, Brines

1. Introduction

Soil sullying with organic and inorganic waste products is attracting attention due to toxic effects on natural vegetation, wildlife and human health [1]. Kerosene, diesel and petrol (petroleum hydrocarbons) are powerful (potent) soil organic pollutants [2] [3] [4]. Petroleum contamination could be due to spillage, leakage, discriminating disposal of petroleum products, road traffic, industrial seepage and agricultural activities [2] [5] [6] [7] [8] [9]. Petroleum of which kerosene and petrol are derivatives contain sulfur, nitrogen and oxygen in low concentrations as well as metals such as lead, nickel, sodium, calcium, copper, uranium and manganese [10] [11] [12]. It can cause chronic or acute effects on the plants. In addition to direct and indirect toxicity, the oil causes interference in the hydric relations of the plants. The toxicity leads to a reduction in the growth of the stem and the root [13].

[14] reported that toxic hydrocarbon molecules could inhibit the activities of amylase and starch phosphorylase and thereby affecting the assimilation of starch. The most common and important symptoms observed in the plants contaminated with oil and its byproducts is a degradation of chlorophyll [15] [16] [17] alterations in the stomatal mechanism [15], reduction in photosynthesis and respiration [15].

Oil spillage has been known to exhibit various deleterious effects on plants by generally retarding plant growth [18] [19] [20]; reduces aeration, by blocking air spaces between soil particles, hence creates conditions of anaerobiosis [21], and causes root stress which reduces leaf growth thereby impairing photosynthesis [15] [16] [17] [22].

Plants may be killed by oil pollution or suffer reduced growth and reproductive rates due to a combination of physical coating, altered soil chemistry and toxic effects of crude oil components [5]. Also, different heavy metals at supra optimal concentrations have been shown to inhibit various metabolic processes in plants resulting in their reduced growth and development [11] [23].

Sodium chloride (brine) when introduced in the soil could increase soil salinity and thereby affects vegetation in the surrounding area. The crop growth may be altered by the interaction between salinity and nutrients in the soil [24]. [25] reported the adverse effect of a flow in roadside salinity on irrigated agriculture.

Electrical resistivity, a physical property of the soil, can be affected by these pollutants and is determined by standard measurement of soil resistance by measuring the distance and cross sectional area through which current travels. Artificially generated electric currents are supplied to the soil and the resulting potential differences are measured. Potential difference patterns provide information in the form of subsurface heterogeneities and of their electrical properties [26]. This study is focused on investigation of the resistivity and other properties of polluted soils and the consequent effect on the growth rate of plants.

2. Method

A method is presented for constructing an inexpensive plug in apparatus for measuring the electrical resistivity of polluted soil samples in the laboratory. The principle of operation of the apparatus is based on Ohm's law. A 12v dry cell, voltmeter, ammeter was connected together with the constructed pipe and soil was introduced into the pipe as shown in **Figure 1**. Electric current is injected into soil in the pipe using two conducting metal electrodes connected to 12v battery as a power source. The potential differences ΔV between two points A and B and flow of current I in a linear line with the current electrode were measured by the voltmeter and ammeter respectively.

From Ohm's law the current, which flows through a conductor of length, L and cross sectional area S, is directly proportional to the potential difference across the ends of the conductor that is:

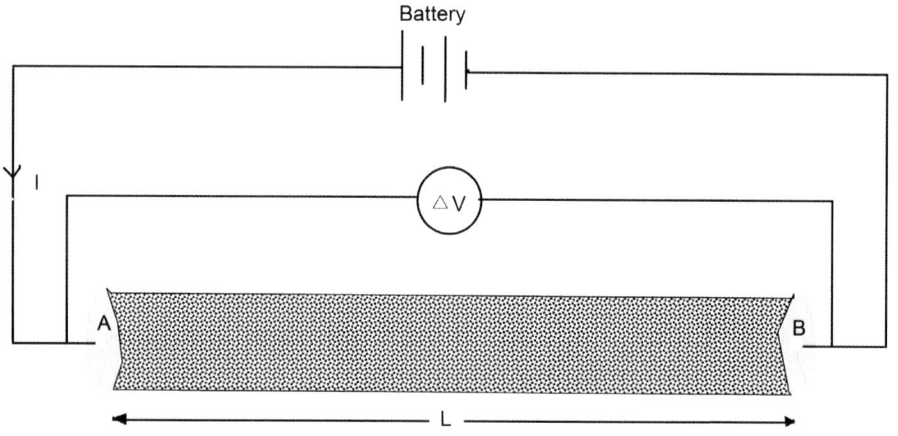

Figure 1. Experimental set up for determining the resistivity of contaminated soil in the laboratory.

$$\Delta V = IR$$

where R = resistance of the conductor (soil). It follows that

$$R = \frac{\Delta V}{I} \tag{1}$$

and R is also expressed as

$$R = \frac{\rho L}{S} = \frac{\Delta V}{I}$$

$$\rho = \left(\frac{\Delta V}{I}\right)\left(\frac{S}{L}\right) \tag{2}$$

ρ is the resistivity of the soil to be determined.

From the values of voltage (V) and current (A) measured, the resistivity is determined.

Soil was introduced into the pipes of various lengths and compacted. The initial resistivity of the soil was determined before introducing the contaminants into the soil. The contaminants introduced into the soil in the various pipes included kerosene, petrol and brine (salt solution) of different concentrations. The concentrations of brine solution are 1.0, 0.5, 0.33 and 0.10 molar. The contaminated soil samples were left for 24 hours for proper transmission into the pore spaces. Each pipe containing the soil or contaminated soil samples was connected in turns as shown in **Figure 1** and current and voltage measured. The resistivity was therefore calculated using Equation (2).

The contaminants were tested on germinated beans plants to know the effective rate of growth. Six nursery bags were filled with planting soil and beans were sown in them. After germination, the plants were allowed to grow for some days. The growth rate was then measured and recorded each day for another five (5) days. Contaminants (Petrol, Kerosene and Various salt solution concentrations) were introduced into five nursery plants and one was left to serve as a control. The height of the plants from the stem above the soil level to the tip of the apical bud (terminal bud) was measured at a specified time each day. The growth rates were recorded for another five (5) consecutive days. Precaution was taken so that the contaminant was not applied directly on the plants.

3. Results and Discussion

Figure 2 showed the resistivity of the soil when different quantities of kerosene were added. It was observed that resistivity increases with volume in the three different pipes after absorption. **Figure 3** equally showed the resistivity of soil to be directly proportional to the volume of petrol added. Addition of different volumes of salt solution showed inverse variation to the resistivity (see **Figure 4**).

Figure 5 showed that resistivity was also inversely proportional to the length of pipe used when the same volume and concentration of salt solution was added to the soil. **Figure 6** showed that when the concentration of the salt solution was increased, the resistivity of the soil sample reduced.

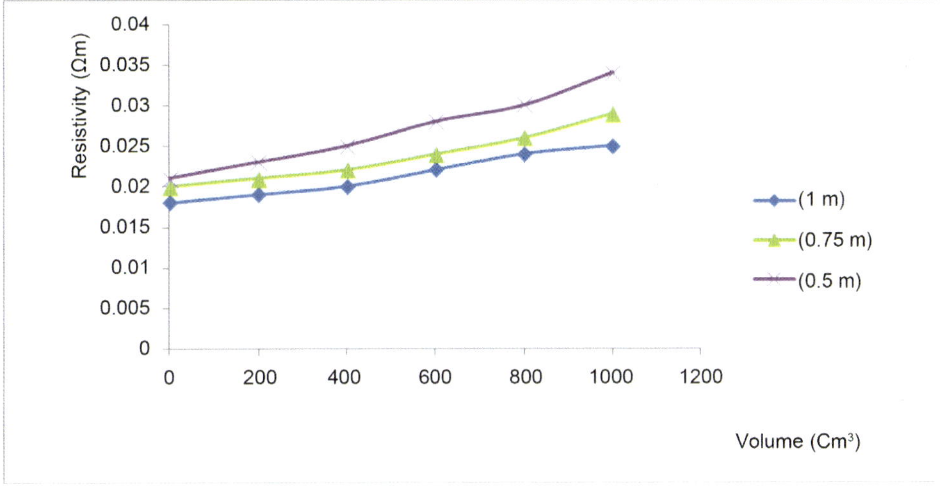

Figure 2. Graph of resistivity against the volume of kerosene added for different lengths (0.5 m, 0.75 m and 1.0 m).

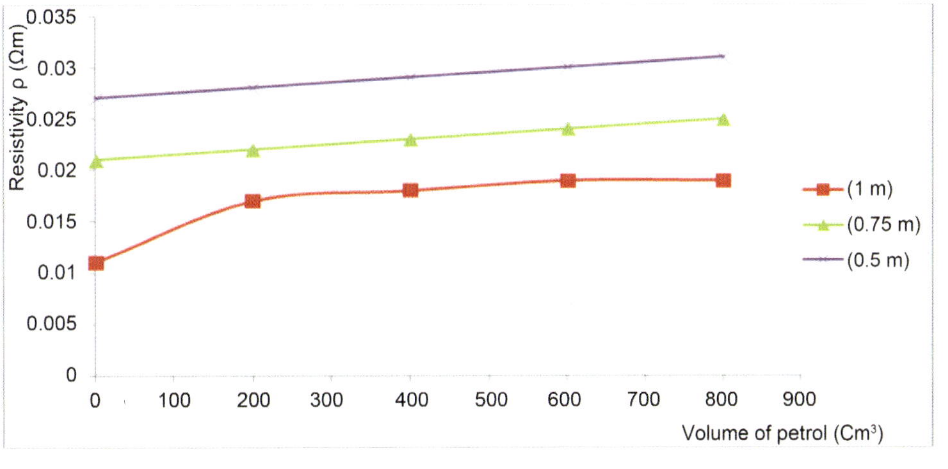

Figure 3. Graph of resistivity against the volume of petrol added for different lengths (0.5 m, 0.75 m and 1.0 m).

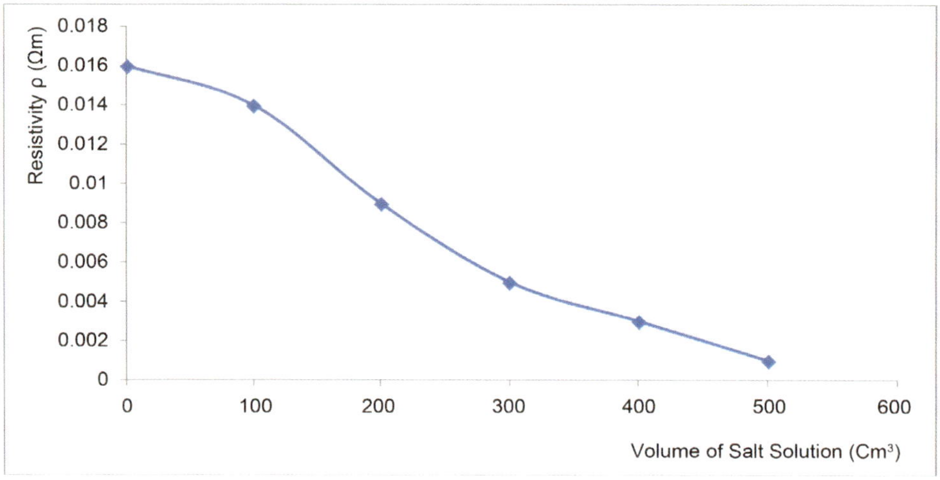

Figure 4. Adding brine solution to soil sample (at constant length of pipe) decreases its resistivity.

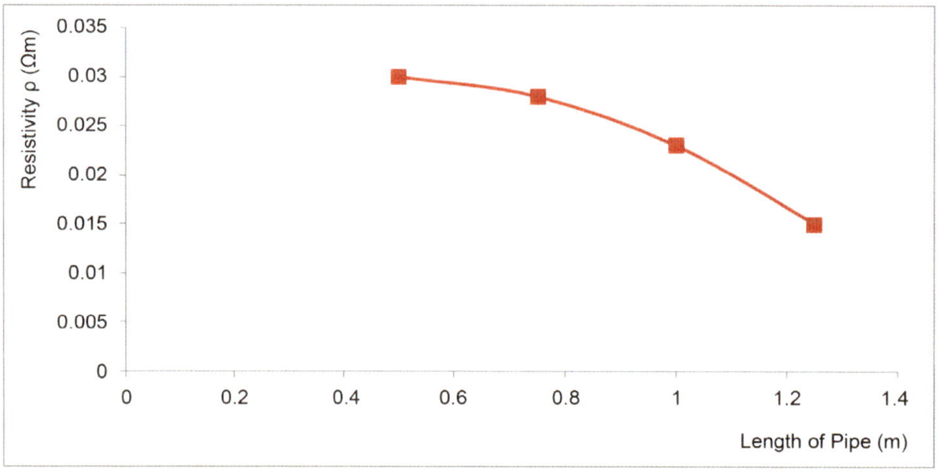

Figure 5. Graph of resistivity versus lengths (0.5 m, 0.75 m, 1.0 m and 1.25 m) of pipe at constant volume and concentration of salt solution.

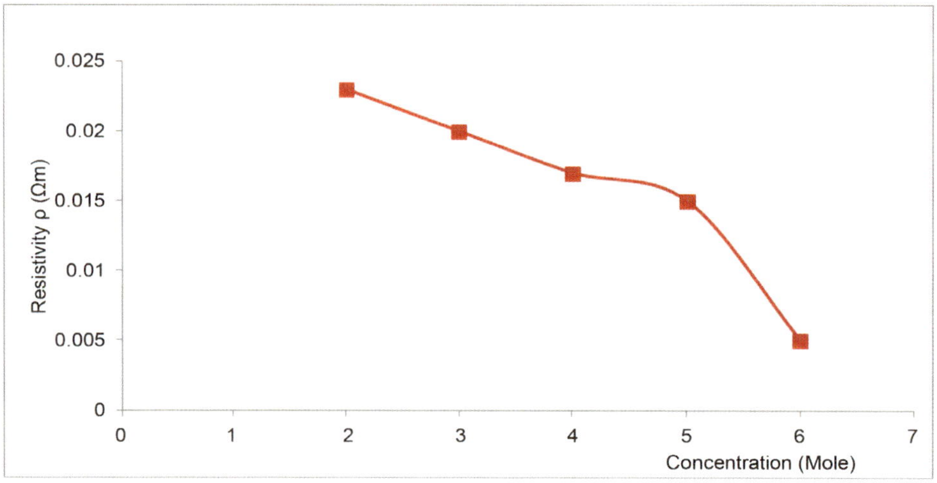

Figure 6. Resistivity is inversely proportional to the concentration of salt solution (at constant length of pipe).

In essence, pollutants such as kerosene, petrol, salt solution have effects on the resistivity of soil. From the result obtained above, it is evident that petrol and kerosene increase the resistivity of the soil while the increase in length of the tube and concentration of salt solution cause decrease in the soil resistivity, thereby increasing the soil conductivity. Generally, salt solution contains ions and when introduced into the soil increases the conductivity of the soil as is seen in **Figure 4**. Also increase in concentration of the salt solution increases the salinity of the soil, and thereby lower the resistivity, **Figure 6**.

From **Figure 7** it was also noticed that there was no growth, that is increase in height from the stem above soil level to the tip of the apical bud, in nursery bean plants D, E and F after Day 6 and by Day 11 the plants were dead. Hence, the death of the plants in nursery bean plants B, C and D with different salt concentrations was in agreement [27], that excess salinity in soil can decrease plant available water and cause plant stress.

Plants may also be killed by oil pollution or suffer reduced growth and reproductive rates due to a combination of physical coating, altered soil chemistry and toxic effects of crude oil components [5].

The nursery bean plant C was in bending position, this could be due to high concentration of salt while the plants in D, E and F later withered and died. The control A, showed a steady and normal growth in height, **Figure 7**. Results obtained from this study agreed with [28] that used engine oil as a contaminant to study its effect on plant height, stem girth and moisture content. A number of researchers had revealed that crude oil inhibits plant growth [29], reduces germination due to toxic effects on seeds [30] and leads to decrease in biomass productivity [31].

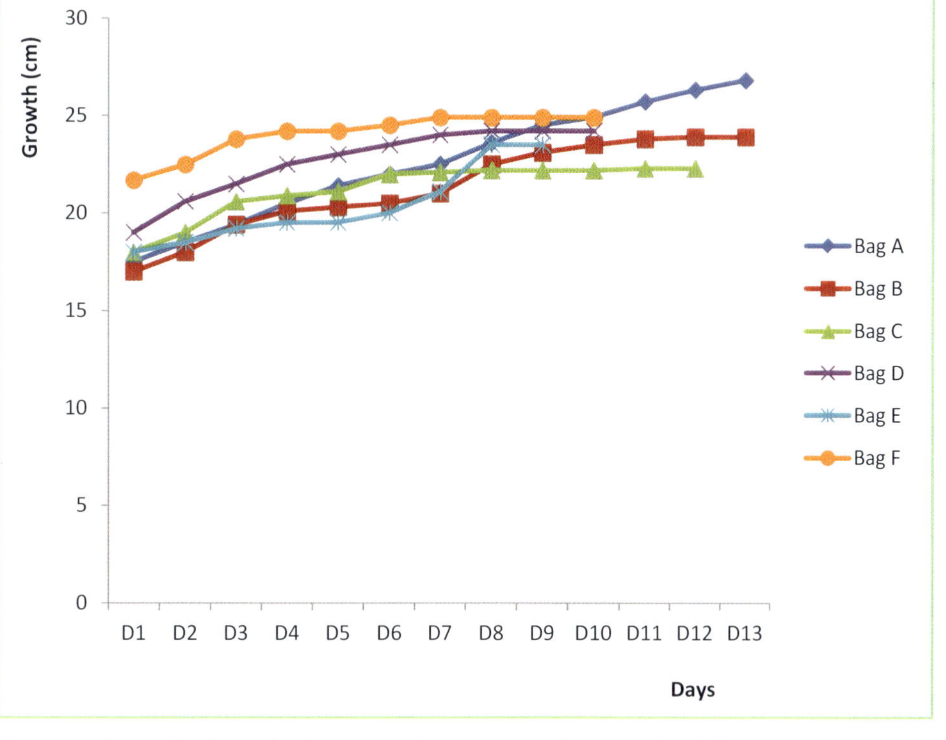

Figure 7. The graph of growth of germinated beans against days in soil of different contaminants.

4. Conclusions

The effect of the brine solution showed that, the higher the concentration of salt, the lower is the resistivity and for petrol and kerosene the higher the volume, the higher the resistivity. Length on the other hand is inversely proportional to the resistivity.

The study also showed that Kerosene, Petrol and Salt solution of different concentrations have an adverse effect on the growth and development of the plants and also on physical properties of the soil. The presence of salinity in soils and other contaminates is detrimental to plant growth.

References

[1] Rahbar, F.G., Kiarostami, K. and Shirdam, R. (2012) Effects of Petroleum Hydrocarbons on Growth, Photosynthetic Pigments and Carbohydrate Levels of Sunflower. *Journal of Food, Agriculture and Environment*, **10**, 773-776.

[2] Alkorta, I. and Garbisu, C. (2001) Phytoremediation of Organic Contaminants in Soils. *Bioresource Technology*, **79**, 273-276. http://dx.doi.org/10.1016/S0960-8524(01)00016-5

[3] Villalobos, M., Avila-Forcada, A.P. and Gutierrez-Ruiz, M.E. (2008) An Improved Gravimetric Method to Determine Total Petroleum Hydrocarbons in Contaminated Soils. *Water, Air and Soil Pollution*, **194**, 151-161. http://dx.doi.org/10.1007/s11270-008-9704-1

[4] Asghar, H.N., Setia, R. and Marschner, P. (2012) Community Composition and Activity of Microbes from Saline Soils and Non-Saline Soils Respond Similarly to Changes in Salinity. *Soil Biology and Biochemistry*, **47**, 175-178. http://dx.doi.org/10.1016/j.soilbio.2012.01.002

[5] Leighton, F.A. (2000) Petroleum oils and Wildlife. CCWHC Wild Health Topic, 345 p.

[6] Vasudevan, N. and Rajaram, P. (2001) Bioremediation of Oil Sludge-Contaminated Soil. *Environment International*, **26**, 409-411. http://dx.doi.org/10.1016/S0160-4120(01)00020-4

[7] Peng, S., Zhoua, Q., Cai, Z. and Zhang, Z. (2009) Phytoremediation of Petroleum Contaminated Soils by *Mirabilis jalapa* L. in a Greenhouse Plot Experiment. *Journal of Hazardous Materials*, **168**, 1490-1496. http://dx.doi.org/10.1016/j.jhazmat.2009.03.036

[8] Ahmadu, J. and Egbodion, J. (2013) Effect of oil spillage on cassava production in Niger Delta region of Nigeria. *American Journal of Experimental Agriculture*, **3**, 914-926. http://dx.doi.org/10.9734/AJEA/2013/4374

[9] Agbogidi, O.M. and Arinze, A.N. (2014) Effects of Kerosene Contaminated Soil on the Growth of Soya Bean (*Glycine max* (L.) Merr.). *International Journal of Environmental Science*, **2**, 54-61.

[10] Posthuma, J. (1970) The Composition of Petroleum. *Rapports et Proces-verbaux des Réunions. Conseil International pour l'Éxploration de la Mer.*, **171**, 7-16.

[11] Davies, E. (1991) Effects of Toxin Concentration of Metals on Root Growth and Development. In: Artkinson, D., Ed., *Plant Root Growth*, Blackwell, London, 211-227.

[12] Archambault, D.J. and Winterhalder, W.K. (1995) Metal Tolerance in Agrostic Scabra Ontario Area. *Canadian Journal of Botany*, **78**, 766-775. http://dx.doi.org/10.1139/b95-084

[13] Alkio, M., Tabuchi, T.M., Wang, X. and Colón-Carmona, A. (2005) Stress Responses to Polycyclic Aromatic Hydrocarbons in Arabidopsis Include Growth Inhibition and Hypersensitive Response-Like Symptoms. *Journal of Experimental Botany*, **56**, 2983-2994. http://dx.doi.org/10.1093/jxb/eri295

[14] Achuba, F.I. (2006) The Effect of Sublethal Concentrations of Crude Oil on the Growth and Metabolism of Cowpea (*Vigna unguiculata*) Seedlings. *Environmentalist*, **26**, 17-20.

http://dx.doi.org/10.1007/s10669-006-5354-2

[15] Baker, J.M. (1970) The Effects of Oils on Plants. *Environmental Pollution*, **1**, 27-44. http://dx.doi.org/10.1016/0013-9327(70)90004-2

[16] Malallah, G., Afzal, M., Kurian, M., Gulshan, S. and Dhami, M.S.I. (1998) Impact of Oil Pollution on Some Desert Plants. *Environment International*, **24**, 919-924. http://dx.doi.org/10.1016/S0160-4120(98)00061-0

[17] Malallah, G., Afzal, M., Gulshan, S., Kurian, M. and Dhami, M.S.I. (1996) *Vicia faba* as a Bioindicator of Oil Pollution. *Environmental Pollution*, **92**, 213-217. http://dx.doi.org/10.1016/0269-7491(95)00085-2

[18] Gill, L.S. and Sandate, R.M.A. (1974) Effect of Foliarly Applied CCC on the Growth of *Phaseolus aureus*. Rjexb-mung on Green Gram. *Bangladesh Journal of Biological Sciences*, **15**, 35-40.

[19] Glouse, A.K.M., Zaidi, H. and Attique, A. (1980) Effect of Air Pollution on the Foliar Organs of *Callistemon citrinus Staf. Journal of Scientific Research*, **2**, 207-209.

[20] Ekpo, M.A. and Nwankpa, L.I. (2005) Effect of Crude on Microorganisms and Growth of Ginger (*Zingba officinate*) in the Tropics. *Journal of Sustainable Tropical Agricultural Research*, **16**, 67-71.

[21] Rowell, M.J. (1977) The Effects of Crude Oil Spill on Soils. A Review of Literature 1, The Reclamation of Agricultural Soil after Oil Spills, Part 1. Department of Soil Science, University of Alberta Canada, Canada, 33-355.

[22] Smith, B., Stachowish, M. and Volken-Burgh, E. (1989) Cellular Processes Limiting Growth in Plants under Hypoxic Root Stress. *Experimental Botany*, **40**, 89-94. http://dx.doi.org/10.1093/jxb/40.1.89

[23] Bala, R. and Setia, R.C. (1990) Some Aspects of Lead Toxicity in Plants. Narendia Publishing House, India, 268.

[24] Fores, P., Carvajal, M., Cerda, A. and Marrinez, V. (2001) Salinity and Ammonium/Nitrate Interactions on Tomato Plant Development, Nutrition and Metabolites. *Journal of Plant Nutrition*, **24**, 1561-1573. http://dx.doi.org/10.1081/PLN-100106021

[25] Chinnusamy, N., Jagendorf, A. and Zhu, J. (2005) Understanding and Improving Salt Tolerance in Plants. *Crop Science*, **45**, 437-448. http://dx.doi.org/10.2135/cropsci2005.0437

[26] Kearey, P., Brooks, M. and Hill, I. (2002) An Introduction to Geophysical Exploration. Blackwell Science, Oxford.

[27] Krista, P.E. (2003) The Basics of Salinity and Sodicity Effects on Soil Physical Properties (Information Highlight for the General Public). Adapted from a paper by Nikos J. Warrence, Krista E. Pearson and James W. Bauder. http://waterquality.montana.edu/docs/methane/basics_highlight.shtml

[28] Adenipekun, C.O. and Kassim, L.O. (2006) Effects of Spent Engine Oil on the Growth Parameters and Moisture Content of *Celosia argentea*. In: In: Akpan, G. and Odoemena, C.S.J., Eds., *Botany and Environmental Health*, University of Uyo, Uyo, 108-111.

[29] Cook, F.D. and Westlake, D.W.S. (1974) Information Canada. Cat. No R72-12774.

[30] Udo, E.J. and Opara, O.O. (1984) Some Studies on the Effects of Crude Oil Pollution of Soil on Plant Growth. *Journal of Biology and Applied Chemistry*, **11**, 26-29.

[31] Odejimi, R.A.O. and Ogbalu, O. (2006) Physiological Impact of Crude Oil Polluted Soil on Growth, Carbohydrate and Protein Levels of Edible Shoot of Fluted Pumpkin (Telfera occidentalis). In: Akpan, G. and Odoemena, C.S.J., Eds., *Botany and Environmental Health*, University of Uyo, Uyo, 102-105.

7

Pollution Potential Ranking of Industries using Classical TODIM Method

Nisha Soni, Robin A. Christian, Namrata Jariwala

Civil Engineering Department, SVNIT, Surat, India
Email: nps2412@gmail.com, rac@ced.svnit.ac.in, ndj@ced.svnit.ac.in

Abstract

Pollution of air and water is increasing day by day, as the number of industry is increasing and sometimes emission of pollutants in air and water exceeds the norms. So the pollution potential of industries based on the emission of pollutants needs to be assessed. In proposed method, based on the weights and emission of the pollutants, the final score of industries has been calculated which indicates the polluting potential of the industry. In this paper, attempt has been made to evaluate the score and the rank of industries using TODIM decision making process, in which higher score of particular industry is an indicator of higher pollution load imparted by particular industry and vice versa.

Keywords

MADM, Pollution Potential, Classical TODIM

1. Introduction

In developing country, the development of nation depends on industrial growth. The numbers of industries are increasing, which will help in economical growth of nation. If we look other side, increasing in numbers of industries will also increase the pollution of air, water and soil. We should think about the emission from the different industries. There are a number of pollutants emitted by different types of industries in surrounding environment. It may happen that the emission of pollutants will exceed the norm provided by government. For some typical industry, it may happen that out of total pollution load, only some pollutants emission may exceed norms but others are within the prescribed limits. It may be the case in the industrial sector that primary industry may not emit any harmful pollutants in the environment but compared to those small scale or secondary industries, emission load will be closed to norms but more harmful.

If the pollutant load will be more than prescribed limit, it will be always considered in the category of polluting industries. It is very difficult to judge the potentiality of pollutants in terms of different emission load. In proposed method, impact of each pollutant treated separately and later on ranking of industry will be calculated after combining impact of all pollutants together for each industry.

The sections in this paper are arranged as follows. Section 2 describes environment pollution associated with the industries. Section 3 gives brief introduction of the classical TODIM method. In Section 4, two examples are evaluated with the help of the proposed method. Lastly, the conclusion of this paper is presented in Section 5.

2. Environment Pollution

Industrial growth is required for every nation in this 21st century to be with recent technology and also for human comfort. Industrial growth is directly connected with environment pollution. As the growth in industries increasing rapidly the pollution of environment also increase in proportion. Before few decades there were less number of industries and the working hours of industries were also less. In present scenario industries are increasing in number and majority of industries are operated for 10 - 12 hours in day. Finally all these affect environment pollution. The issue of industrial pollution has taken on grave importance for agencies trying to fight against environmental degradation. Nations facing sudden and rapid growth of such industries are finding it to be a serious problem which has to be brought under control immediately.

Pollution is mainly related to air, water and land. Surrounding environment of any industry is depending on emission of that industry. Some industries may pollute air and water more compare to land. In this paper we are considering few industries which are only responsible for air and water pollution emission. To analyze proposed method the data has been collected data of three chemical industries and three thermal power plant units from consultant who are doing the regular monitoring of the industries for environmental audit purpose.

There are many pollutants emitted from above said industries. Among all those considerations have been given to pollutants which are more hazardous to living things. Pollutants like SO_x, NO_x, SPM and Cl_2 emitted in air and SS, TDS, COD, BOD, Chlorides, Sulphates, Phosphates, Oil & Grease, Total Chromium and Total Iron emitted in water are more harmful to living things are considered for calculating the pollution potential.

Proposed method is used to rank all such industries which described in next section, and accordingly government can take the strict action against this type of industries.

3. The Classical TODIM Method

There are number of methods by which we can solve the multiple attribute problems [1] [2] [3] [4]. Almost all methods used expected utility theory to solve MADM problems, in which the strict assumption of the decision maker (DM) is required regarding complete rationality, but behavioral experimental of DM is bounded in such decision process [5]. So in such a case DM's psychological behavior is used to solve the MADM problem in ongoing research topic.

Many useful multiple attribute methods are available to researchers in decision aiding at present. These methods are based on different mathematical assumptions. A number of accomplishments have been achieved since the late 1970s and new developments and application fields are constantly emerging. Behavioral decision theory is considered to have formally begun with Ward Edwards in Psychological Bulletin, although it went through some major advances in the 1970s and 1980s [6] [7] [8].

TODIM (Interactive and Multiple Attribute Decision Making, acronym in Portuguese) was first introduced by Gomes and Lima. Later on it was applied on many practical MADM problems, like residential properties evaluation, selection of the destination of natural gas and so on [9] [10] [11] [12]. The classical TODIM method, based on Prospect Theory, is used to select best alternative among all available alternatives. In this method the dominance degree of each alternative over the others are measured first, which reflects DM's behavioral characteristic like reference dependence. Then after by calculating the overall score of each alternative ranking can be assign to each alternative. TODIM is used to solve the MADM problem by considering DM's behavior.

The main idea behind the classical TODIM is to calculate the dominance degree of each alternative over all the others alternatives by considering a multi-attribute function based on Prospect Theory [13] [14] [15] [16] [17]. The ranking of alternatives will be observed by this obtained dominance degree. The classical TODIM method used to solve only MADM problem which crisp numbers. The TODIM method does not deal with risk directly, the way the decision maker evaluates the outcomes of any decision can be expressed by their risk attitude: for instance, a cautious decision maker will undervalue a superior result more than a braver one. The algorithm for the TODIM method is described as follows.

Step 1: Decision matrix

$\boldsymbol{X} = \left[x_{ij}\right]_{m\times n}$ is normalized into $Y_j = \left[y_{ij}\right]_{m\times n}$ by normalizing method

where x_{ij} = crisp number, $i \in M$, $j \in N$.

Many methods are available for normalizing the matrix. But here we use the following method for normalizing the decision matrix.

$$y_{ij} = x_{ij} \Big/ \sum_{i=1}^{m} x_{ij} \,. \tag{1}$$

Step 2: Now relative weight w_{jr} of each attribute C_j calculated by taking C_r as a reference attribute.

$$w_{jr} = w_j / w_r \,, \quad j, r \in N \tag{2}$$

where $w_r = \max\left\{w_j \mid j \in N\right\}$.

Step 3: Now the dominance degree of alternative A_i over alternative A_k is calculated by concerning attribute C_j.

$$\Phi(A_i, A_k) = \begin{cases} \sqrt{\dfrac{(y_{ij} - y_{kj}) w_{jr}}{\left(\sum_{j=1}^{n} w_{jr}\right)}}, & (y_{ij} - y_{kj}) > 0 \\ 0, & (y_{ij} - y_{kj}) = 0 \\ \left(-\dfrac{1}{\theta}\right) \sqrt{\dfrac{(y_{ij} - y_{kj})\left(\sum_{j=1}^{n} w_{jr}\right)}{w_{jr}}}, & (y_{kj} - y_{ij}) > 0 \end{cases} \quad (3)$$

where θ = losses attenuation factor.

$y_{ij} - y_{kj}$ = gain of alternative A_i over alternative A_k concerning attribute C_j if $y_{ij} - y_{kj} > 0$ and the loss if $y_{ij} - y_{kj} < 0$.

θ stands for the degree of loss. The greater θ is, the lower the degree of loss is. Because of that sensitivity analysis should applied to affirm the stability of the calculated results by considering decision makers' preferences. So sensitivity analysis should carry out on θ, on the criteria weights, on the choice of the reference criterion, and on performance evaluations.

Apart from parameter θ, the attenuation factor of the losses, function Φc does not offer other parameters to delineate the behavior of diverse decision makers, therefore a generic formulation is proposed.

Step 4: Now the overall dominance degree of alternative A_i over alternative A_k is calculated.

$$\delta(A_i, A_k) = \sum_{j=1}^{n} \Phi_j(A_i, A_k),\ i, k \in M. \quad (4)$$

Step 5: Now the overall value of alternative A_i is calculated.

$$\xi(A_i) = \frac{\sum_{k=1}^{m} \delta(A_i, A_k) - \min\left\{\sum_{k=1}^{m} \delta(A_i, A_k)\right\}}{\max\left\{\sum_{k=1}^{m} \delta(A_i, A_k)\right\} - \min\left\{\sum_{k=1}^{m} \delta(A_i, A_k)\right\}},\ i \in M. \quad (5)$$

Step 6: Now give rank to all alternatives and select the most suitable one(s) according to the overall scores of alternatives. The higher $\xi(A_i)$ the better alternative A_i.

Obviously value of $\xi(A_i)$ is between 0 and 1 ($0 \le \xi(A_i) \le 1$). Therefore arranging overall values of all the alternatives in descending order, we can obtain the ranking of given alternatives or we can select the most suitable alternative(s).

4. Numerical Examples

In this section, two examples are used to explore the use of the extended TODIM method and finally to rank industries.

Example 1. On the basis of weight established for the sub criteria, the Environmental Pollution Potential Hazard Index was developed for three chemical industries having monitored emission data are taken for study analysis. The three industries are Industry 1, Industry 2, and Industry 3. Here analysis of data is divided in two parts. 1) Score using attributes of air 2) score using attributes of water. Using the TODIM method the calculations are carried out in following manner.

Score using attributes of air. First decision matrix is formed and is shown in **Table 1**. Here the attributes are the emission of SO_X, NO_X, SPM and Cl_2. Increase in the values of the attributes will increase the air pollution potential. Attribute weight vector provided by the DM is $\omega = (0.222, 0.203, 0.221, 0.187)$. To rank the industries, the method proposed in this paper is used and the procedure is summarized below.

Step 1: The decision matrix is normalized using Equation (1) and is shown following **Table 2**.

Step 2: Relative weights of each attributes are calculated using Equation (2) and shown below.

$$w_{jr} = \{1.0000, 0.9144, 0.9955, 0.8423\}.$$

Here, in the calculation process, $w_r = \max\{0.222, 0.203, 0.221, 0.187\} = 0.222$.

Step 3: Using Equation (3) the dominance degree matrices concerning attributes SO_X, NO_X, SPM and Cl_2 are constructed, respectively, *i.e.*

$$\Phi_1 = \begin{bmatrix} 0.0000 & -3.2115 & -1.6122 \\ 0.1900 & 0.0000 & 0.1643 \\ 0.0954 & -2.7775 & 0.0000 \end{bmatrix}, \quad \Phi_2 = \begin{bmatrix} 0.0000 & -2.0290 & 0.0201 \\ 0.1098 & 0.0000 & 0.1116 \\ -0.3717 & -2.0627 & 0.0000 \end{bmatrix}$$

$$\Phi_3 = \begin{bmatrix} 0.0000 & -1.0073 & -2.4639 \\ 0.0593 & 0.0000 & -2.2486 \\ 0.1451 & 0.1324 & 0.0000 \end{bmatrix}, \quad \Phi_4 = \begin{bmatrix} 0.0000 & 0.2232 & 0.2232 \\ -4.4795 & 0.0000 & 0.0000 \\ -4.4795 & 0.0000 & 0.0000 \end{bmatrix}$$

Here $\theta = 1$ is taken which indicates that the losses are equally contribute with their real value to the global value [13].

Step 4: Using Equation (4) the overall dominance degree matrix δ built, *i.e.*

$$\delta = \begin{bmatrix} 0.0000 & -6.0245 & -3.8328 \\ -4.1204 & 0.0000 & -1.9727 \\ -4.6106 & -4.7078 & 0.0000 \end{bmatrix}.$$

Table 1. The decision matrix.

Alternatives	Attributes			
	SO_X	NO_X	SPM	Cl_2
Industry 1	9.59	29.1667	65	4.0233
Industry 2	83.9667	54	85	0
Industry 3	28.3333	28.3333	184.6667	0

Table 2. The normalized decision matrix.

Alternatives	Attributes			
	SO_X	NO_X	SPM	Cl_2
Industry 1	0.07867	0.2615	0.1942	1
Industry 2	0.6888	0.4843	0.2539	0
Industry 3	0.2324	0.2541	0.5517	0

Step 5: Using Equation (5) the overall value of each alternative can be obtained, *i.e.*, ξ(industry 1) = 0, ξ(industry 1) = 1, ξ(industry 3) = 0.1431.

Step 6: According to the overall values, the ranking order of the three industries is determined, industry 2 > industry 3 > industry 1. Obviously, industry 2 is the most desirable industry.

***Score using attributes of water*:** First decision matrix is formed and is shown in **Table 3**. Here the attributes are the emission of SS, TDS, COD, BOD, Chlorides, Sulphates, Phosphates and Bio-assay. Increase in the values of the attributes will increase the water pollution potential. Attribute weight vector provided by the DM is $w = (0.119, 0.135, 0.144, 0.149, 0.123, 0.095, 0.109, 0.125)$. To rank the industries, the method proposed in this paper is used and the procedure is summarized below.

Step 1: The decision matrix is normalized using Equations (1) and (2) is shown in **Table 4**.

Step 2: Relative weights of each attributes are calculated using Equation (2) and shown below.

$$w_{jr} = \{0.7987, 0.9060, 0.9664, 1.0000, 0.8255, 0.6376, 0.7315, 0.8389\}.$$

Here, in the calculation process,

$w_r = \max\{0.119, 0.135, 0.144, 0.149, 0.123, 0.095, 0.109, 0.125\} = 0.149$.

Step 3: Using Equation (3) the dominance degree matrices concerning attributes SS, TDS, COD, BOD, Chlorides, Sulphates, Phosphates and Bio-assay are constructed, respectively, *i.e.*

$$\Phi_1 = \begin{bmatrix} 0.0000 & 0.0445 & 0.0754 \\ -2.5064 & 0.0000 & 0.0609 \\ -4.2498 & -3.4320 & 0.0000 \end{bmatrix}, \quad \Phi_2 = \begin{bmatrix} 0.0000 & 0.0708 & 0.0905 \\ -3.5182 & 0.0000 & 0.0564 \\ -4.4967 & -2.8004 & 0.0000 \end{bmatrix}$$

$$\Phi_3 = \begin{bmatrix} 0.0000 & -1.9533 & 0.0406 \\ 0.0420 & 0.0000 & 0.0584 \\ -1.8913 & -2.7189 & 0.0000 \end{bmatrix}, \quad \Phi_4 = \begin{bmatrix} 0.0000 & -2.1617 & 0.0434 \\ 0.0480 & 0.0000 & 0.0647 \\ -1.9513 & -2.9121 & 0.0000 \end{bmatrix}$$

Table 3. The decision matrix.

Alternatives	Attributes							
	SS	TDS	COD	BOD	Chlorides	Sulphates	Phosphates	Bio-assay
Industry 1	91.333	4500	86.3333	28.3333	375.287	1525	0	0.7
Industry 2	70	2468.667	107.6667	37.3333	121.6667	0.5867	5.3167	0.6167
Industry 3	30	1181.667	66.3333	21	366.6667	541.6667	0	0.9833

Table 4. The normalized decision matrix.

Alternatives	Attributes							
	SS	TDS	COD	BOD	Chlorides	Sulphates	Phosphates	Bio-assay
Industry 1	0.4774	0.5521	0.3316	0.3269	0.4346	0.7377	0.0000	0.3043
Industry 2	0.3659	0.3029	0.4136	0.4308	0.1409	0.0003	1.0000	0.2681
Industry 3	0.1568	0.1450	0.2548	0.2423	0.4246	0.2620	0.0000	0.4275

$$\Phi_5 = \begin{bmatrix} 0.0000 & 0.0734 & 0.0135 \\ -4.0010 & 0.0000 & -3.9324 \\ -0.7376 & 0.0721 & 0.0000 \end{bmatrix}, \ \Phi_6 = \begin{bmatrix} 0.0000 & 0.1022 & 0.0821 \\ -7.2141 & 0.0000 & -4.2980 \\ -5.7940 & 0.0609 & 0.0000 \end{bmatrix}$$

$$\Phi_7 = \begin{bmatrix} 0.0000 & -7.8429 & 0 \\ 0.1275 & 0.0000 & 0.1275 \\ 0.0000 & -7.8429 & 0.0000 \end{bmatrix}, \ \Phi_8 = \begin{bmatrix} 0.0000 & 0.0260 & -2.5704 \\ -1.3938 & 0.0000 & -2.9239 \\ 0.0479 & 0.0545 & 0.0000 \end{bmatrix}$$

Here $\theta = 1$ is considered.

Step 4: Using Equation (4) the overall dominance degree matrix δ built, *i.e.*

$$\delta = \begin{bmatrix} 0.0000 & -11.6410 & -2.2248 \\ -18.4160 & 0.0000 & -10.7864 \\ -19.0728 & -19.5188 & 0.0000 \end{bmatrix}.$$

Step 5: Using Equation (5) the overall value of each alternative can be obtained, *i.e.*, ξ (industry 1) = 1.0000, ξ(industry 1) = 0.3797, ξ(industry 3) = 0.

Step 6: According to the overall values, the ranking order of the three industries is determined, industry 1 > industry 2 > industry 3. Obviously, industry 1 is the most desirable industry.

Here the attributes of air and water are calculated separately. But finally ranking of any industry will be combination of both, attributes of air and water. And then we can get final score. Based on that final score we can give the rank to industries. And so it is calculated and shown in **Table 5**.

It can be seen from **Table 6** that the ranking orders of alternatives are:

Example 2. On the basis of weight established for the sub criteria, the Environmental Pollution Potential Hazard Index was developed for three Thermal Power Station Units having monitored emission data are taken for study analysis. The three Thermal Power Station Units are Unit 1, Unit 2, and Unit 3. Here analysis of data is divided in two parts. 1) Score using attributes of air 2) score using attributes of water. Using the TODIM method the calculations are carrying out in following manner.

Score using attributes of air: First decision matrix is formed and is shown in **Table 7**.

Here the attributes are the emission of SO_X, NO_X, and SPM. Increase in the values of the attributes will increase the air pollution potential. Attribute weight vector provided

Table 5. Overall Score of attributes of air and water.

Criteria	Industry 1	Industry 2	Industry 3	Weight
Air Pollution	0.0000	1.0000	0.1431	0.5
Water Pollution	1.0000	0.3796	0.0000	0.5
Overall Pollution	0.5000	0.6898	0.0715	

Table 6. Ranking of industries

Chemical Industries	Rank
1	2
2	1
3	3

Table 7. The decision matrix.

Alternatives	Attributes		
	SO_X	NO_X	SPM
Unit 1	7.36	8.0333	3.6
Unit 2	9.18	137.3333	3.0333
Unit 3	5.0333	7.5	0

by the DM is $w = (0.329, 0.317, 0.355)$. To rank the industries, the method proposed in this paper is used to rank thermal power station units.

Step 1: The decision matrix is normalized using Equations (1) and (2) is shown in **Table 8**.

Step 2: Relative weights of each attributes are calculated using Equation (2) and shown below.

$$w_{jr} = \{0.9268, 0.8930, 1.0000\}.$$

Here, in the calculation process, $w_r = \max\{0.329, 0.317, 0.355\} = 0.355$.

Step 3: Using Equation (3) the dominance degree matrices concerning attributes SO_X, NO_X, and SPM are constructed respectively, *i.e.*

$$\Phi_1 = \begin{bmatrix} 0.0000 & -0.8503 & 0.1122 \\ 0.0992 & 0.0000 & 0.1498 \\ -0.9614 & -1.2835 & 0.0000 \end{bmatrix}, \quad \Phi_2 = \begin{bmatrix} 0.0000 & -2.7429 & 0.0198 \\ 0.3084 & 0.0000 & 0.3090 \\ -0.1762 & -2.7486 & 0.0000 \end{bmatrix}$$

$$\Phi_3 = \begin{bmatrix} 0.0000 & 0.1037 & 0.2614 \\ -0.8238 & 0.0000 & 0.2399 \\ -2.0762 & -1.9058 & 0.0000 \end{bmatrix}.$$

Step 5: Using Equation (5) the overall value of each alternative can be obtained, *i.e.*, $\xi(\text{industry 1}) = 0.6419$, $\xi(\text{industry 1}) = 1$, $\xi(\text{industry 3}) = 0$.

Step 6: According to the overall values, the ranking order of the three industries is determined, industry 2 > industry 1 > industry 3. Obviously, industry 2 is the most desirable industry.

Score using attributes of water: First decision matrix is formed and is shown in **Table 9**.

Here SS, TDS, BOD, Phosphates, Oil & Grease, Total Chromium, Total Iron are attributes. Attribute weight vector provided by the DM is $w = (0.151, 0.118, 0.209, 0.135, 0.113, 0.134, 0.140)$. To rank the industries, the method proposed in this paper is used to rank thermal power station.

Step 1: The decision matrix is normalized using Equation (1) and is shown in **Table 10**.

Step 2: Relative weights of each attributes are calculated using Equation (2) and shown below.

$$w_{jr} = \{0.7225, 0.8660, 1.0000, 0.6459, 0.5407, 0.6411, 0.6699\}$$

Here, in the calculation process,

Table 8. The normalized decision matrix.

Alternatives	Attributes		
	SO_X	NO_X	SPM
Unit 1	0.3412	0.0524	0.5427
Unit 2	0.4255	0.8986	0.4573
Unit 3	0.2333	0.0491	0.0000

Table 9. The decision matrix.

Alternatives	Attributes						
	SS	TDS	BOD	Phosphates	Oil & Grease	Total Chromium	Total Iron
Unit 1	26.6667	780.6667	15.8	1.6833	0.4667	0.0127	0.0287
Unit 2	47	912.6667	13.3333	1.14333	0	0.0143	0.036
Unit 3	30.1333	838.6667	0	0.746	0	0.0127	0.0313

Table 10. The normalized decision matrix.

Alternatives	Attributes						
	SS	TDS	BOD	Phosphates	Oil & Grease	Total Chromium	Total Iron
Unit 1	0.2569	0.3083	0.5423	0.4712	1.0000	0.3199	0.2990
Unit 2	0.4528	0.3605	0.4577	0.3200	0.0000	0.3602	0.3750
Unit 3	0.2903	0.3312	0.0000	0.2088	0.0000	0.3199	0.3260

$$w_r = \max\{0.151, 0.118, 0.209, 0.135, 0.113, 0.134, 0.140\} = 0.209 .$$

Step 3: Using Equation (3) the dominance degree matrices concerning attributes SS, TDS, BOD, Phosphates, Oil & Grease, Total Chromium and Total Iron are constructed respectively, *i.e.*

$$\Phi_1 = \begin{bmatrix} 0.0000 & -2.4914 & -1.0287 \\ 0.0786 & 0.0000 & 0.0716 \\ 0.0325 & -2.2691 & 0.0000 \end{bmatrix}, \quad \Phi_2 = \begin{bmatrix} 0.0000 & -1.4539 & -0.9638 \\ 0.0359 & 0.0000 & 0.0268 \\ 0.0238 & -1.0886 & 0.0000 \end{bmatrix}$$

$$\Phi_3 = \begin{bmatrix} 0.0000 & 0.0608 & 0.1539 \\ -1.3922 & 0.0000 & 0.1414 \\ -3.5236 & -3.2369 & 0.0000 \end{bmatrix}, \quad \Phi_4 = \begin{bmatrix} 0.0000 & 0.0653 & 0.0860 \\ -2.3145 & 0.0000 & 0.0560 \\ -3.0493 & -1.9853 & 0.0000 \end{bmatrix}$$

$$\Phi_5 = \begin{bmatrix} 0.0000 & 0.1537 & 0.1537 \\ -6.5071 & 0.0000 & 0.0000 \\ -6.5071 & 0.0000 & 0.0000 \end{bmatrix}, \quad \Phi_6 = \begin{bmatrix} 0.0000 & -1.1996 & 0.0000 \\ 0.0336 & 0.0000 & 0.0336 \\ 0.0000 & -1.1996 & 0.0000 \end{bmatrix}$$

$$\Phi_7 = \begin{bmatrix} 0.0000 & -1.6121 & -0.9621 \\ 0.0472 & 0.0000 & 0.0378 \\ 0.0282 & -1.2935 & 0.0000 \end{bmatrix}.$$

Here $\theta = 1$ is considered.

Step 4: Using Equation (4) the overall dominance degree matrix δ built, *i.e.*

$$\delta = \begin{bmatrix} 0.0000 & -6.4772 & -2.5609 \\ -10.0186 & 0.0000 & 0.3673 \\ -12.9957 & -11.0730 & 0.0000 \end{bmatrix}.$$

Step 5: Using Equation (5) the overall value of each alternative can be obtained, *i.e.*, ξ (industry 1) = 1.0000, ξ(industry 2) = 0.9592, ξ(industry 3) = 0.

Step 6: According to the overall values, the ranking order of the three industries is determined, industry 1 > industry 2 > industry 3. Obviously, industry 1 is the most desirable industry.

Here the attributes of air and water are calculated separately. But finally ranking of any industry will be the combination of both, attributes of air and water. And then we can get final score. Based on that final score we can give the rank to industries. The final score of industries are calculated and shown in **Table 11**. And then rank is assigned to all industries.

It can be seen from **Table 12** that the ranking orders of alternatives are

5. Conclusions

The whole study demonstrates the use of classical TODIM method for the ranking of the industries based on their appraisal score, here Pollution Potential, with the case studies. The classical TODIM method, used in this study, can effectively capture Decision Maker's psychological behavior, to solve this type of problems under hesitant fuzzy environment. Following conclusion has been derived from the case studies.

As the pollution level is increasing, it is suggested that the issue of pollution tax must be studied and evaluated by behavior of decision makers of developing countries to control the pollution levels in the environment. However in this method, all method all attributes considered as equally. It means whether attribute is benefit/beneficial or non-benefit/cost/non-beneficial, all will be treat same. This limitation has impact on considering this method in area of different attributes which include both categories.

Table 11. The overall score.

Criteria	Industry 1	Industry 2	Industry 3	Weight
Air Pollution	0.6419	1.0000	0.0000	0.5
Water Pollution	1.0000	0.9592	0.0000	0.5
Overall Pollution	0.5148	1.0000	0.0000	

Table 12. Ranking of 3 Thermal Power Station Units

Thermal Power Station Units	Rank
1	2
2	1
3	3

Acknowledgements

This work was partly supported by Prof. Paresh Solanki, Ex. Student, ECED, SVNIT, Surat.

References

[1] Lourenzutti, R. and Krohling, R.A. (2015) TODIM Based Method to Process Heterogeneous Information, Information Technology and Quantitative Management (ITQM). *Procedia Computer Science*, **55**, 318-327. http://dx.doi.org/10.1016/j.procs.2015.07.056

[2] de Souza, T.T.M. and Krohling, R.A. (2012) Fuzzy TODIM for Group Decision Making. *Proceedings of CLAIO SBPO*, Rio de Janeiro, Brazil, 24-28 September 2012, 466-474.

[3] Zhang, J.J., Wu, D.S. and Olson, D.L. (2005) The Method of Grey Related Analysis to Multiple Attribute Decision Making Problems with Interval Numbers. *Mathematical and Computer Modelling*, **42**, 991-998. http://dx.doi.org/10.1016/j.mcm.2005.03.003

[4] Fazlollahtabar, H., Eslami, H. and Salmani, H. (2010) Designing a Fuzzy Expert System to Evaluate Alternatives in Fuzzy Analytic Hierarchy Process. *Journal of Software Engineering and Applications*, **3**, 409-418. http://dx.doi.org/10.4236/jsea.2010.34046

[5] Tversky, A. and Kahneman, D. (1992) Advances in Prospect Theory: Cumulative Representation of Uncertainty. *Journal of Risk and Uncertainty*, **5**, 297-323. http://dx.doi.org/10.1007/BF00122574

[6] Edwards, W. (1954) The Theory of Decision Making. *Journal of Psychological Bulletin*, **51**, 380-417. http://dx.doi.org/10.1037/h0053870

[7] Slovlc, P., Flschhoff, B. and Lichtenstein, S. (1977) Behavioural Decision Theory. *Journal of Annual Review of Psychology*, **28**, 1-39. http://dx.doi.org/10.1146/annurev.ps.28.020177.000245

[8] Gomes, L.F.A.M. and Gonzalez, X.I. (2012) Behavioural Multi-Criteria Decision Analysis: Further Elaborations on the TODIM Method. *Foundations of Computing and Decision Sciences*, **37**, No. 1.

[9] Gomes, L.F.A.M., Rangel, L.A.D. and Maranhao, F.J.C. (2009) Multicriteria Analysis of Natural Gas Destination in Brazil: An Application Of The TODIM Method. *Mathematical and Computer Modelling*, **50**, 92-100. http://dx.doi.org/10.1016/j.mcm.2009.02.013

[10] Gomes, L.F.A.M., Gomes, C.F.S. and Rangel, L.A.D. (2009) A Comparative Decision Analysis with Thor and TODIM: Rental Evaluation in Volta Redonda. *Journal of Revista Tecnologia Fortaleza*, **30**, 7-11.

[11] Rangel, L.A.D., Gomes, L.F.A.M. and Cardoso, F.P. (2011) An Application of the TODIM Method to the Evaluation of Broadband Internet Plans. *Journal of Pesquisa Operacional*, **31**, 235-249. http://dx.doi.org/10.1590/S0101-74382011000200003

[12] Krohling, R.A. and de Souza, T.T.M. (2012) F-TODIM: An Application of the Fuzzy TODIM Method to Rental Evaluation of Residential Properties. *Proceedings of CLAIO SBPO*, Rio de Janeiro, Brazil, 24-28 September 2012, 431-443.

[13] Ramooshjan, K., Rahmani, J., Sobhanollahi, M.A. and Mirzazadeh, A. (2015) A New Method in the Location Problem using Fuzzy TODIM. *Journal of Human and Social Science Research*, **6**, 001-013.

[14] Pereira, J., Gomes, L.F.A.M., Paredes, F., Fuenzalida, R. and Olguin, N. (2014) A Multicriteria Analysis of Five Ordering Methods Including the Bullwhip Effect Criterion. *Proceedings of Recent Advances in Theory, Methods and Practice of Operation Research*, Monterrey, Mexico, October 2014, 239-246.

[15] Rangel, L.A.D., Gomes, L.F.A.M. and Moreira, R.A. (2009) Decision Theory with Multiple Criteria: An Application of ELECTRE IV and TODIM to SEBRAE/RJ. *Journal of Pesquisa Operacional*, **29**, 577-590. http://dx.doi.org/10.1590/S0101-74382009000300007

[16] Abdellaoui, M., Bleichrodt, H. and Paraschiv, C. (2007) Loss Aversion under Prospect Theory: A Parameter-Free Measurement. *Journal of Management Science*, **53**, 1659-1674. http://dx.doi.org/10.1287/mnsc.1070.0711

[17] Kahneman, D. and Tversky, A. (1979) Prospect Theory: An Analysis of Decision under Risk. *Journal of Econometrica*, **47**, 263-292. http://dx.doi.org/10.2307/1914185

8

Morphology-Controlled Synthesis of $Zn_xCd_{1-x}S$ Solid Solutions: An Efficient Solar Light Active Photocatalyst for the Degradation of 2,4,6-Trichlorophenol

Rengaraj Selvaraj[1,2*], Kezhen Qi[3], Sathish Babu Soundra Pandian[4], Mohammed A. Meetani[5], Bushra Al Wahaibi[1], Haider Al Lawati[1], Salma M. Z. Al Kindy[1], Younghun Kim[6], Mika Sillanpää[2]

Department of Chemistry, College of Science, Sultan Qaboos University, Muscat, Sultanate of Oman
Laboratory of Green Chemistry, LUT Savo Sustainable Technology, Lappeenranta University of Technology, Mikkeli, Finland
Institute of Catalysis for Energy and Environment, College of Chemistry and Chemical Engineering, Shenyang Normal University, Shenyang, China
CAARU, College of Science, Sultan Qaboos University, Muscat, Sultanate of Oman
Department of Chemistry, United Arab Emirates University, Al Ain, UAE
Department of Chemical Engineering, Kwangwoon University, Seoul, South Korea
Email: *srengaraj1971@yahoo.com

Abstract

$Zn_xCd_{1-x}S$ solid solutions with controlled morphology have been successfully synthesized by a facile solution-phase method. The prepared samples were characterized by X-ray powder diffraction (XRD), UV-vis diffuse reflectance spectra, X-ray photoelectron spectroscopy (XPS), scanning electron microscopy (SEM) and transmission electron microscopy (TEM). The photocatalytic activity of $Zn_xCd_{1-x}S$ was evaluated in the 2,4,6-trichlorophenol (TCP) degradation and mineralization in aqueous solution under direct solar light illumination. The experiment demonstrated that TCP was effectively degraded by more than 95% with 120 min. The results show that ZnS with Cd doping ($Zn_xCd_{1-x}S$) exhibits the much stronger visible light adsorption than that of pure ZnS, the light adsorption increasing as the Cd^{2+} doping amount. These results indicate that Cd doping into a ZnS crystal lattice can result in the shift of the valence band of ZnS to a positive direction. It may lead to its higher oxidative ability than pure ZnS, which is important for organic pollutant degradation under solar light irradiation. Furthermore, the photocatalytic activity studies reveal that the prepared $Zn_xCd_{1-x}S$ nanostructures exhibit an excellent photocatalytic performance, degrading rapidly the aqueous 2,4,6-trichlorophenol solution under solar light irradiation. These results suggest that $Zn_xCd_{1-x}S$ nanostructure will be a promising candidate of photocatalyst working in solar light range.

Keywords

Photocatalyst, Nanostructures, Solution Phase, 2,4,6-Trichlorophenol, Solar Light, Photocatalytic Activity

1. Introduction

Among the water contaminants, chlorophenols are the most hazardous class of water pollutants [1]. Chlorophenols (CPs) are used in many industries such as paint, pharmaceutical, pesticide, solvent, wood, paper, and pulp industries [2]. However, these CPs are very harmful to humans because they are toxic, mutagenic and carcinogenic. They affect the nervous system and cause respiratory problems [3]. They must be decomposed before discharging to the rivers or lakes. Under this background, it is of importance to find an easy and effective method to completely remove these chlorinated organic pollutants [4].

During recent decades, semiconductor based photocatalysis has been extensively studied and has become one of the most promising methods in environmental protection procedures such as air and water purification and hazardous waste remediation [5]. TiO_2 nanomaterials have often been used as photocatalysts because of their efficiency, stability, non-toxicity, and low cost [6]. However, it has some drawbacks such as the relatively high value of the band gap energy (~3.2 eV), which limits its absorption to the UV spectral region, and the high recombination rate of photo-induced electrons and holes, which decreases its photocatalytic activity [7]. Therefore, many efforts have focused on investigating high-efficiency visible-light photocatalysts because of their potential for various applications, including environmental pollution control and solar energy utilization.

Metal sulfides have been intensively studied as active photocatalysts because of their high light harvesting for promoting charge separation [8]-[12]. For example, CdS shows a high photocatalytic activity in water splitting for hydrogen production [13]. However, there remain some drawbacks including fast recombination of photogenerated electron-hole pairs and suffer from severe photocorrosion, both of which prohibit the wide application of CdS. To solve these problems, forming CdS/ZnS heterojunction structure is an effective method, because the same coordination mode between CdS and ZnS will promote interfacial charge transfer and inhibit the photo-corrosion of CdS via dispersing ZnS on the surface as well [14]. The key point of this CdS/ZnS heterojunction is to achieve a strong join between CdS and ZnS intersurface and the uniform distribution of each kind of atoms inside bulk material. Moreover, the electronic structure of the $Zn_xCd_{1-x}S$ solid solution can be easily controlled by tuning the molar ratio of Zn^{2+} and Cd^{2+} [15]. Therefore, the synthesis of $Zn_{1-x}Cd_xS$ solid solutions is of immense interest and increasing technological importance for future applications.

The controlled synthesis of inorganic nanomaterials with desired morphology has received considerable attention due to the physical and chemical properties of sold nanomaterials specially depend on their phase, size, shape, and organization [16]-[25]. Over the past few years, many methods have been developed to synthesize $Zn_xCd_{1-x}S$ solid solutions with controlled morphology, such as solid state reactions [26], solution-phase process [27], cation exchange reactions [28], and microwave synthesis [29]. However, it is still a big challenge to find an easy way to prepare $Zn_xCd_{1-x}S$ solid solutions with controlled morphology. Also unfortunately, there has been no report available for the degradation of 2,4,6-trichlorophenol using $Zn_xCd_{1-x}S$ solid solutions. In this paper, we report for the first time on the photocatalytic degradation of 2,4,6-trichlorophenol (TCP). The result shows that, with the increase of Cd molar fraction, the character of UV-visible absorption of $Zn_xCd_{1-x}S$ crystal gradually evolves from ZnS to CdS, and this $Zn_xCd_{1-x}S$ sample under solar irradiation demonstrates a considerable photocatalytic degradation of aqueous 2,4,6–trichlorophenol (TCP).

2. Experimental

2.1. Synthesis of $Zn_xCd_{1-x}S$ Samples

The $Zn_xCd_{1-x}S$ solid solutions with different Cd molar ratios were prepared with $Cd(NO_3)_2 \cdot 4H_2O$, $Zn(NO_3)_2 \cdot 6H_2O$ and thioacetamide. The $Zn_xCd_{1-x}S$ solid solutions were prepared according to the previous reported methods [26] [27]. All chemicals were purchased as guaranteed regents and used without further purification. The stock solution was prepared by dissolving amount of $Cd(NO_3)_2 \cdot 4H_2O$, $Zn(NO_3)_2 \cdot 6H_2O$ and thioacetamide in a 250 mL round bottom flask and by filling with 100 mL of distilled water. Afterwards, the stock solution was heated to reflux at 105˚C. After the mixture was refluxed for 45 min, a precipitate was obtained, which was filtered then washed with absolute ethanol and distilled water several times. After being dried in oven at 60˚C for 24 hours, the products were collected for characterization.

2.2. Characterization

The crystalline properties of Cd doping ZnS nanocrystals were studied by X-ray diffraction (XRD) using a Bruker (D5005) X-ray diffractometer equipped with graphite monochromatized CuK_α radiation (λ = 1.54056 Å). The crystal morphology was examined by field emission scanning electron microscopy (FESEM) and transmission electron microscopy (TEM). The FESEM images were obtained from Hitachi S-4800, which is equipped with an energy dispersive X-ray spectrometer (EDX). The lattice spacing of the products was also taken by high resolution TEM (HRTEM) (JEOL JEM-3010). The chemical states and relative compositions of the samples were studied by X-ray photoelectron spectroscopy (XPS) (ThermoVG, U.K) with a monochromatized Al Kα irradiation (1.4867 eV) and a pass energy of 20 eV. All XPS spectra were obtained with an energy step of 0.1 eV and a dwell time of 50 ms. An Avantage Thermo VG software was used to analyze the XPS data. The absorption spectra of the as-prepared samples were recorded using a UV-Vis diffuse reflectance spectroscopy (DRS) (Jasco V670) in the wavelength range of 200 - 1000 nm, with $BaSO_4$ as a reference.

2.3. Photocatalytic Degradation of 2,4,6-Trichlorophenol

Photocatalytic reactions were performed in a cylindrical borosilicate glass reactor vessel with the volume of 250 mL, with a simulated solar light irradiation for the degradation of 2,4,6-trichlorophenol and a recycling water jacket to keep cooling. The photocatalytic tests were performed with 100 mg of catalyst suspended in 2,4,6-trichlorophenol aqueous solution (250 mL, 5mg/L). The concentration of 2,4,6-trichlorophenol left in the centrifuged aqueous solution was determined by LC/MS spectrophotometer (Model: Agilent Technologies–6460 Triple quad LC/MS), in which Water Symmetry C-8 column (3.5 micro Mt-2.1 × 150–Part No.: WAT106011) was used and mobile phase of methanol/water (95:5 v/v) with 5 mmol ammonium acetate was employed at a flow rate of 0.5 mL/min.

3. Results and Discussion

3.1. Phase Structures

Figure 1 shows the XRD patterns of the as prepared $Zn_xCd_{1-x}S$ (x = 0, 0.2, 0.5, 0.8 and 1) samples. These samples show the similar pattern, and all the diffraction peaks can be well indexed to the cubic ZnS (JCPDS Card No. 75 - 1547). When a small amount of Cd was doped into the ZnS crystal, (for example, $Zn_{0.8}Cd_{0.2}S$), the diffraction peaks of ZnS exhibited an obvious shift toward the lower angle. This implies that Cd^{2+} incorporates into the ZnS crystal lattice and increases the fringe lattice distance of the ZnS crystal due to the larger radius of Cd^{2+} (0.97 Å) than that of the Zn^{2+} ion (0.74 Å) [30]. With

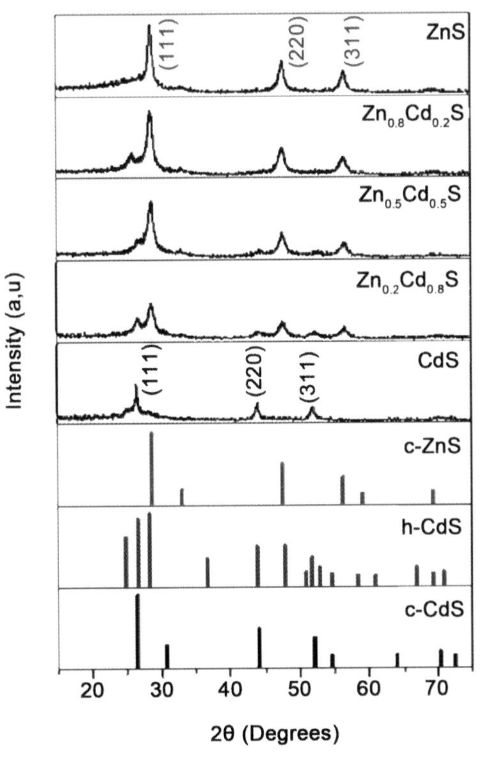

Figure 1. XRD patterns of morphology-controlled $Zn_xCd_{1-x}S$ samples.

increasing Cd^{2+} content in the $Zn_xCd_{1-x}S$ solid solution, the XRD peak positions continuously shift to low angle side, and the crystal phase of the $Zn_xCd_{1-x}S$ solid solution gradually changes from the cubic to hexagonal phase and finally becomes the hexagonal wurtzite CdS phase for the CdS sample (JCPDS Card No. 42-1411). This implies that the Cd^{2+} in the ZnS crystal influence the positions of Zn^{2+} and consequently change the lattice structure of ZnS during the formation of the $Zn_xCd_{1-x}S$ solid solutions.

3.2. UV-Visible Diffuse Reflection Spectra

The optical properties of $Zn_xCd_{1-x}S$ nanostrucures are examined by using UV-vis DRS spectrum, as shown in **Figure 2(a)**. With increasing the molar ratio of Cd^{2+}/Zn^{2+}, the absorption edge of $Zn_xCd_{1-x}S$ shows an obvious red shift. This red shift indicates that the electronic structure of $Zn_xCd_{1-x}S$ nanostrucures can be easily turned by the changing the Zn^{2+}/Cd^{2+} molar ratios in precursors. The samples are changed from white to yellow.

The optical absorption follows the equation áhv = A(hí−E_g)$^{n/2}$, where á, í, E_g and A are the absorption coefficient, the light frequency, the band gap and a constant, respectively. Among them, n determines the characteristics of the transition in a semiconductor [31]. The band gaps are measured to be about 2.22, 2.14 and 2.05 eV for x = 0.2, 0.5 and 0.8 $Zn_xCd_{1-x}S$, respectively (**Figure 2(b)**), which incorporated Cd^{2+} into the ZnS crystal lattice and induced the band gap narrowing.

3.3. XPS Study

In order to study the compositions on $Cd_{0.5}Zn_{0.5}S$ surface, XPS measurements are carried out. **Figure 3(a)** shows the wide spectrum of the $Zn_{0.5}Cd_{0.5}S$ sample, which confirms the presence of Zn, Cd, S, O and C in the sample. The observed C and O peaks is due to the carbon supporting film used for measurement and the adsorbed oxygen molecules on the sample surface, respectively.

Figures 3(b)-(d) shows that the high-resolution XPS spectra of $Zn_{0.5}Cd_{0.5}S$. The peaks at 404.9 (Cd $3d_{5/2}$), 411.7 (Cd $3d_{3/2}$), 1045.5 (Zn $2p_{1/2}$), 1022.3 (Zn $2p_{3/2}$), 163.0 (S $2p_{1/2}$) and 162.0 (S $2p_{3/2}$) were attributed to the $Cd_{0.5}Zn_{0.5}S$ surface condition, which

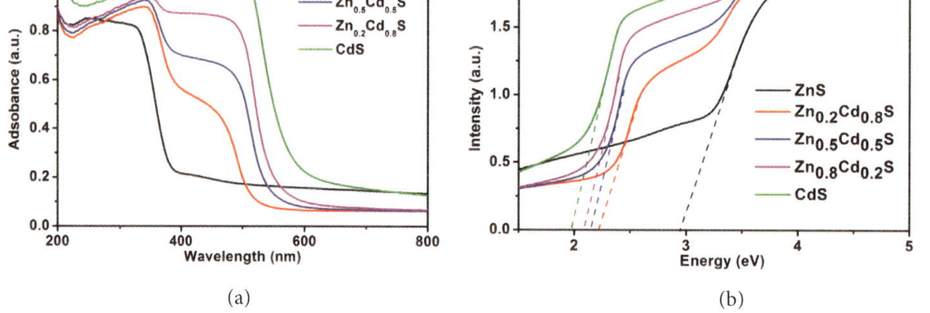

(a) (b)

Figure 2. (a) UV-vis DRS spectra and (b) plot of transformed the energy of the light absorbed for the $Zn_xCd_{1-x}S$ samples.

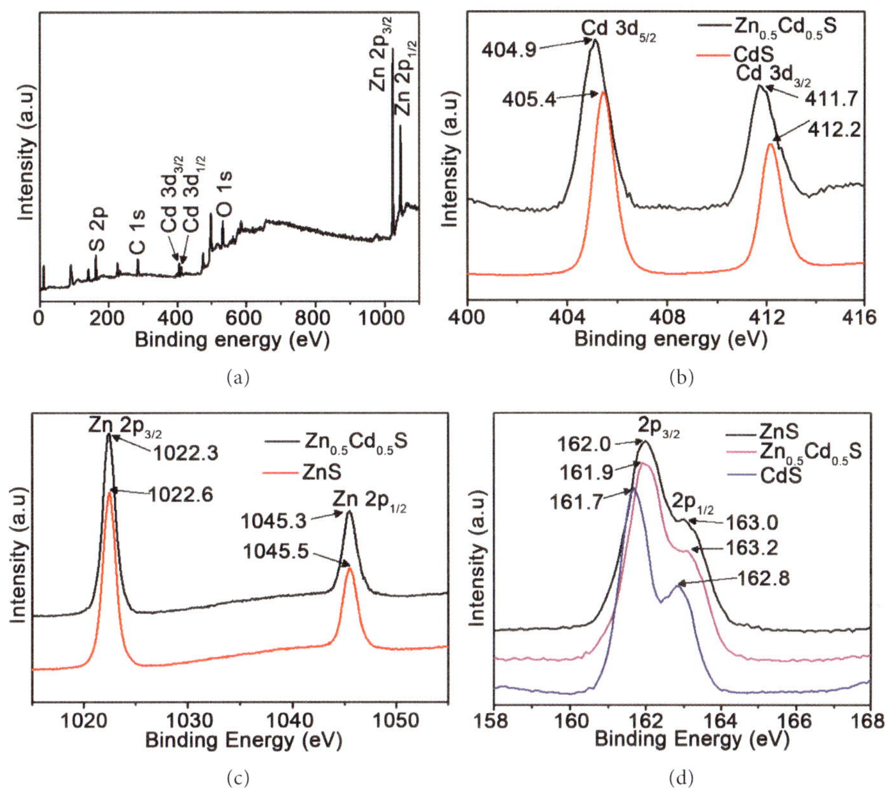

Figure 3. XPS survey spectrum of (a) $Cd_{0.5}Zn_{0.5}S$ and high-resolution XPS spectra of (b) Cd 3d; (c) Zn 2p and (d) S 2p for the $Cd_{0.5}Zn_{0.5}S$ sample.

agree well with the results reported by Yang *et al.* [32] Compare to ZnS, Zn peak of the $Cd_{0.5}Zn_{0.5}S$ show a shift toward low degree (**Figure 3(b)**). Compare to CdS, Cd peak of the $Cd_{0.5}Zn_{0.5}S$ show a shift toward low degree (**Figure 3(c)**). Compare to ZnS and CdS, S peak of the $Cd_{0.5}Zn_{0.5}S$ is in the middle (**Figure 3(d)**).

3.4. SEM Morphology Analysis

SEM measurement was used to investigate the influence of Cd^{2+} doping on the morphology of ZnS nanocrystals, as shown in **Figure 4**. The ZnS sample is composed of the sphere-like nanostructures about 2.5 - 3.0 μm in diameter, the sphere is composed of small nanoparticles (**Figure 4(a)**). When molar ratio of Zn^{2+} to Cd^{2+} is 0.8:0.2, the product is composed of the sphere-like nanostructures illustrated in **Figure 4(b)**, the diameter is about 250 - 300 nm the surface is a little rough. When the molar ratio is 0.5:0.5, the diameters of the spheres are 200 - 300 nm and 80-100 nm, the size is not uniform (**Figure 4(c)**). When the molar ratio is decreased to 0.2:0.8, the diameters of the spheres are in the range of 250 - 270 nm with uniform size (**Figure 4(d)**). The CdS sample is composed of the sphere-like nanostructures about 1.5 - 3.0 μm in diameter, the sphere surface is smooth (**Figure 4(e)**). These SEM results indicate that the general morphology of these $Zn_xCd_{1-x}S$ solid solutions is almost the same, small particles assemble for big particles. However, it is apparent that the diameter of the individual $Zn_xCd_{1-x}S$ spherical particle can be controlled by turning the ratio of Zn^{2+} to Cd^{2+}.

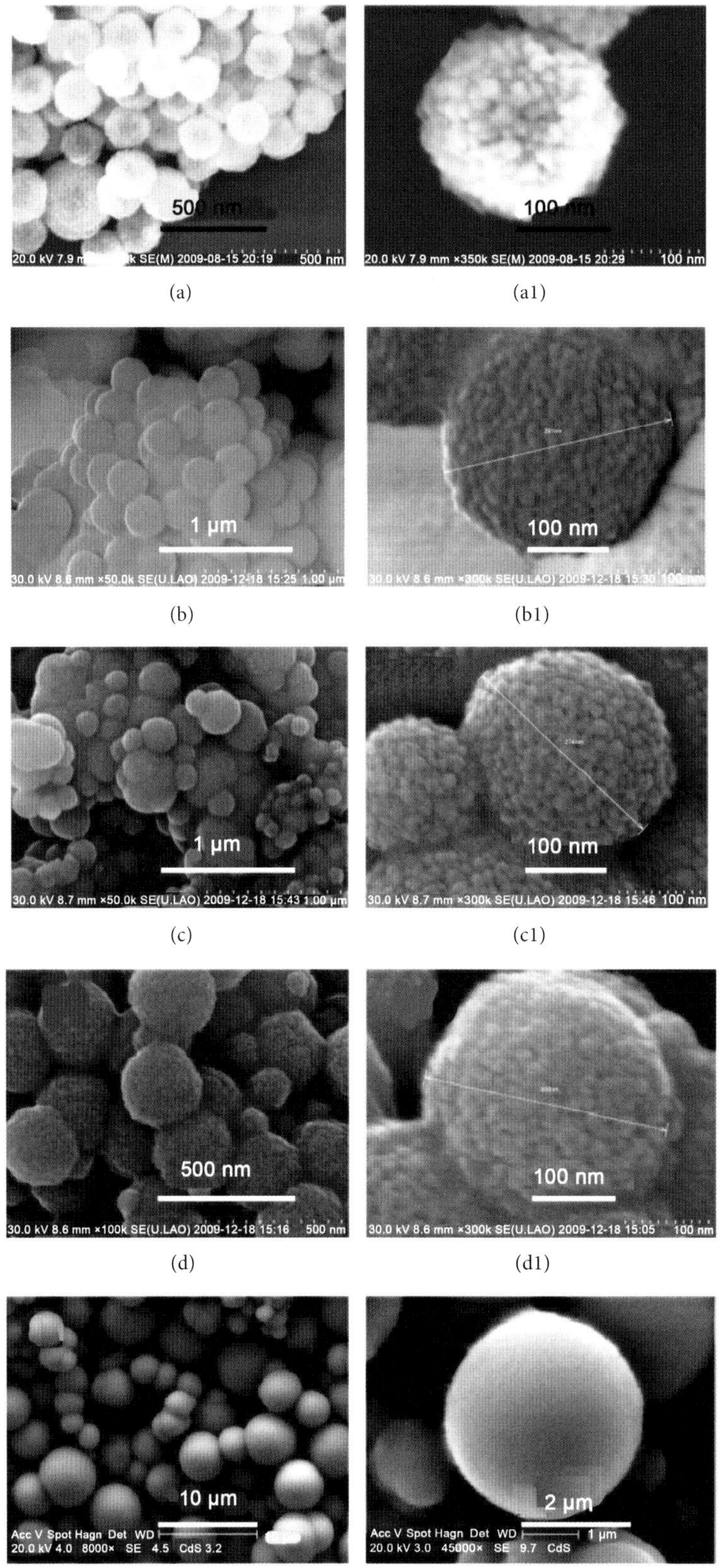

Figure 4. SEM images of the as-prepared $Zn_xCd_{1-x}S$ samples: (a, a1) x = 1, (b, b1) x = 0.8; (c, c1) x = 0.5; (d, d1) x = 0.2; (e, e1) x = 0.

3.5. TEM Test for Samples

The morphologies of as prepared ZnS, $Zn_xCd_{1-x}S$ and CdS samples are investigated using TEM and HRTEM analysis, respectively (**Figure 5**). From the TEM images of as synthesized ZnS samples, it can be seen that the product was present in a large scare spherical with the diameter in 100 - 120 nm (**Figure 5(a)**). There is also spherical particle shape but no uniform size for the as-synthesized $Cd_xZn_{1-x}S$ sample (choose $Zn_{0.5}Cd_{0.5}S$ as an example). Some size is around 50 nm some size is around 300 nm (**Figure 5(b)**). For the CdS sample, make up some irregular nanoparticles with the size of 20-40 nm (**Figure 5(c)**). The diameter of this nanoparticle is consistent with the result from the XRD data according to the calculation of Scherrer Formula. The components of the products were further confirmed by EDX, which indicates that the ratio of Zn, Cd and S is 0.5:0.5:1 for $Zn_{0.5}Cd_{0.5}S$ (**Figures 5(a1)-(c1)**), which is consistent with the XPS result above (**Figure 3**).

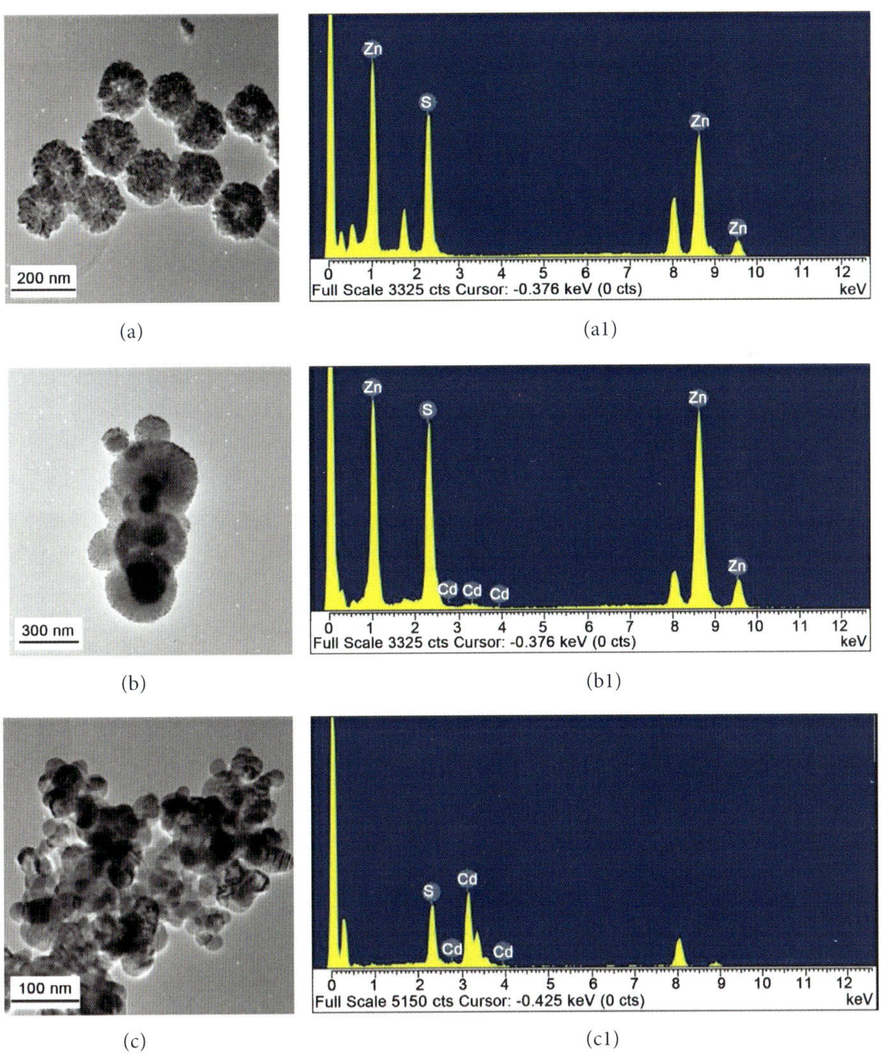

Figure 5. TEM and HRTEM images of as-synthesized (a) ZnS; (b) $Cd_{0.5}Zn_{0.5}S$ and (c) CdS, and the corresponding EDX spectrum.

3.6. Photocatalytic Activity Test

In order to evaluate the photocatalytic activity of $Zn_xCd_{1-x}S$ samples, photodegradation of TCP was carried out in aqueous suspension using pure ZnS and $Zn_xCd_{1-x}S$ catalysts with a different Cd content between 0.2 and 0.8 mole ratio, and the experimental results are shown in **Figure 6**. **Figure 7** represents the time-dependent LC/MS peak intensity of the TCP solution during photo degradation in the presence of CdS microspheres. The peak intensity at around 0.9 min decreased gradually with irradiation time. The experiment demonstrated that TCP was effectively degraded by more than 95% with 120 min. At the end of the degradation after 240 min, no peak could be detected, implying that complete oxidation of TCP occurred due to the presence of CdS microsphere under solar light irradiation. A bigger doping amount of Cd on ZnS can be detrimental to the TCP photodegradation efficiency.

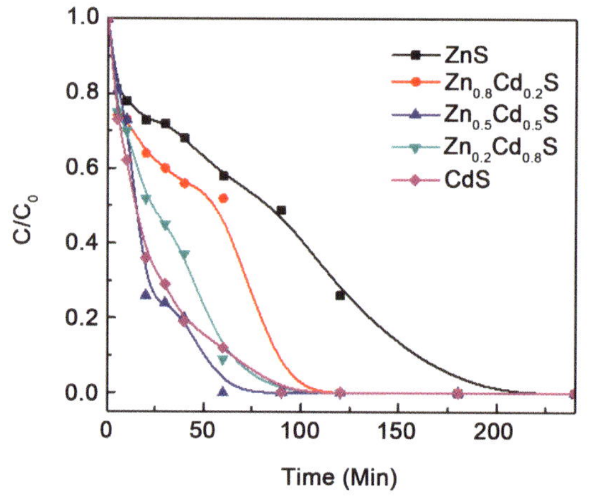

Figure 6. Photocatalytic degradation of 2,4,6-Trichlorophenol under solar light irradiation by $Zn_xCd_{1-x}S$ solid solutions. (Catalyst dosage: 100 mg/250 mL; concentration of 2,4,6-Trichlorophenol: 5 mg/L).

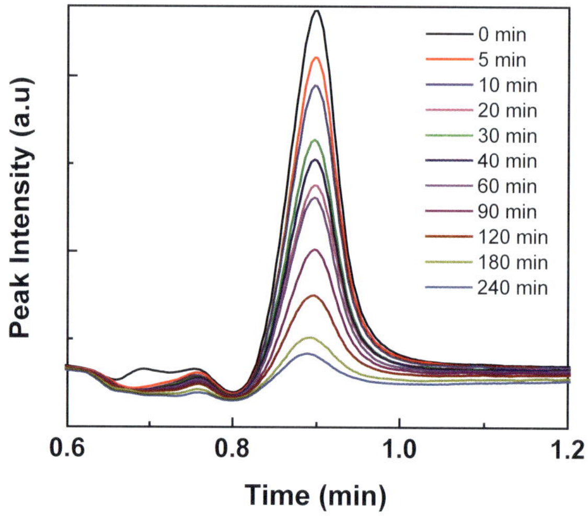

Figure 7. Variation of 2,4,6-trichlorophenol degradation with time by CdS under solar light irradiation.

Figure 6 shows that the photocatalytic activity of $Zn_{0.5}Cd_{0.5}S$ for degradation of TCP is much higher than that of pure ZnS. This result indicates that doping Cd ions promotes the charge pair separation efficiency for ZnS catalysts. The electron activates from the conduction band of ZnS to Cd atoms is thermodynamically due to the Fermi level of ZnS is higher than that of Cd metal [33] [34]. This results in the formation of a Schottky barrier at metal-semiconductor contact region and improves the efficiency of charge separation of ZnS [35] [36]. Contrarily, at the high content of Cd doping, Cd atoms also will act as a recombination center and decrease the photocatalytic activity of ZnS. From this work, the as prepared $Zn_xCd_{1-x}S$ samples demonstrated an excellent performance in photocatalytic degradation of aqueous TCP solution under solar irradiation.

4. Conclusion

In the present study, the feasibility of employing $Zn_xCd_{1-x}S$ solid solutions for the degradation of 2,4,6-trichlorophenol has been demonstrated successfully. $Zn_xCd_{1-x}S$ solid solutions with tunable chemical composition and controlled morphology have been successfully synthesized by a facile solution-phase method without the use of any surfactants. The $Zn_xCd_{1-x}S$ products demonstrate well-dispersed spheres with different diameters. With the increase of Cd molar fractions, the UV-visible absorption of $Zn_xCd_{1-x}S$ crystal gradually evolves from ZnS to CdS, which leads to that the $Zn_xCd_{1-x}S$ sample under solar irradiation demonstrates a considerable photocatalytic degradation of 2,4,6-Trichlorophenol in aqueous solution.

Acknowledgements

The authors (RS and MAM) are grateful to Sultan Qaboos University for providing financial support to carry out this work under SQU-UAEU joint collaborative research grant (CL/SQU-UAEU/16/04).

References

[1] Aslam, M., Soomro, M.T., Ismail, I.M.I., Salah, N., Gondal, M.A. and Hameed, A. (2015) Sunlight Mediated Removal of Chlorophenols over Tungsten Supported ZnO: Electrochemical and Photocatalytic Studies. *Journal of Environmental Chemical Engineering*, **3**, 1901-1911. http://dx.doi.org/10.1016/j.jece.2015.07.004

[2] Abbas, H.A., Jamil, T.S. and Hammad, F.F. (2016) Synthesis, Characterization and Photocatalytic activIty of Nano sized Undoped and Ga Doped $SrTi_{0.7}Fe_{0.3}O_3$ for 2,4,6-Trichlorophenol Photodegradation. *Journal of Environmental Chemical Engineering*, **4**, 2384-2393. http://dx.doi.org/10.1016/j.jece.2016.04.019

[3] Chaliha, S. and Bhattacharya, K.G. (2008) Catalytic Wet Oxidation of 2-Chlorophenol, 2,4-Dichlorophenol and 2,4,6-Trichlorophenol in Water with Mn(II)-MCM41. *Chemical Engineering Journal*, **139**, 575-588. http://dx.doi.org/10.1016/j.cej.2007.09.006

[4] Rengaraj, S. and Li, X.Z. (2006) Enhanced Photocatalytic Activity of TiO_2 by Doping with Ag for Degradation of 2,4,6-Trichlorophenol in Aqueous Suspension. *Journal of Molecular Catalysis A: Chemical*, **243**, 60-67. http://dx.doi.org/10.1016/j.molcata.2005.08.010

[5] Carevic, M., Abazovic, N.D., Savic, T., Novakovic, T.B., Mojovic, M.D. and Comor, M.I. (2016) Structural, Optical and Photodegradation Properties of Pure and Fe-Doped Titania Nanoparticles Probed Using Simulated Solar Light. *Ceramics International*, **42**, 1521-1529. http://dx.doi.org/10.1016/j.ceramint.2015.09.100

[6] Tong, Z., Yang, D., Xiao, T., Tian, Y. and Jiang, Z. (2015) Biomimetic Fabrication of g-C_3N_4/TiO_2 Nanosheets with Enhanced Photocatalytic Activity toward Organic Pollutant Degradation. *Chemical Engineering Journal*, **260**, 117-125. http://dx.doi.org/10.1016/j.cej.2014.08.072

[7] Yang, J., Cui, S.H., Qiao, J. and Lian, H.-Z. (2014) The Photocatalytic Dehalogenation of Chlorophenols and Bromophenols by Cobalt Doped Nano TiO_2. *Journal of Molecular Catalysis A: Chemical*, **395**, 42-51. http://dx.doi.org/10.1016/j.molcata.2014.08.001

[8] Habisreutinge, S.N., Mende, L.S. and Stolarczyk, J.K. (2013) Photocatalytic Reduction of CO_2 on TiO_2 and Other Semiconductors. *Angewandte Chemie International Edition*, **52**, 7372-7408. http://dx.doi.org/10.1002/anie.201207199

[9] Xiang, Q.J., Yu, J.G. and Jaroniec, M. (2012) Synergetic Effect of MoS_2 and Graphene as Cocatalysts for Enhanced Photocatalytic H_2 Production Activity of TiO_2 Nanoparticles. *Journal of the American Chemical Society*, **134**, 6575-6578. http://dx.doi.org/10.1021/ja302846n

[10] Zhang, J., Liu, S.W., Yu, J.G. and Jaroniec, M. (2011) A Simple Cation Exchange Approach to Bi-Doped ZnS Hollow Spheres with Enhanced UV and Visible-Light Photocatalytic H_2-Production Activity. *Journal of Materials Chemistry*, **21**, 14655-14662. http://dx.doi.org/10.1039/c1jm12596f

[11] Li, Q., Guo, B.D., Yu, J.G., Ran, J.R., Zhang, B.H., Yan, H.J. and Gong, J.R. (2011) Highly Efficient Visible-Light-Driven Photocatalytic Hydrogen Production of CdS-Cluster-Decorated Graphene Nanosheets. *Journal of the American Chemical Society*, **133**, 10878-10884. http://dx.doi.org/10.1021/ja2025454

[12] Zhang, J., Qiao, S.Z., Qi, L. and Yu, J. (2013) Fabrication of NiS Modified CdS Nanorod p-n Junction Photocatalysts with Enhanced Visible-Light Photocatalytic H_2-Production Activity. *Physical Chemistry Chemical Physics*, **15**, 12088-12094. http://dx.doi.org/10.1039/c3cp50734c

[13] Xiang, Q.J. and Yu, J.G. (2013) Graphene-Based Photocatalysts for Hydrogen Generation. *The Journal of Physical Chemistry Letters*, **4**, 753-759. http://dx.doi.org/10.1021/jz302048d

[14] Soltani, N., Saiona, E., Yunus, W.M.M., Erfani, M., Navasery, M., Bahmanrokh, G. and Rezaee, K. (2014) Enhancement of Visible Light Photocatalytic Activity of ZnS and CdS Nanoparticles Based on Organic and Inorganic Coating. *Applied Surface Science*, **290**, 440-447. http://dx.doi.org/10.1016/j.apsusc.2013.11.104

[15] Wu, J.C., Zheng, J., Zacherl, C.L., Wu, P., Liu, Z.K. and Xu, R. (2011) Hybrid Functionals Physical Chemistry Chemical PhysicsStudy of Band Bowing, Band Edges and Electronic Structures of $Cd_{1-x}Zn_xS$ Solid Solution. *The Journal of Physical Chemistry C*, **115**, 19741-19748. http://dx.doi.org/10.1021/jp204799q

[16] Zhang, Q., Du, Q., Hua, M., Jiao, T., Gao, F. and Pan, B. (2013) Sorption Enhancement of Lead Ions from Water by Surface Charged Polystyrene-Supported Nano-Zirconium Oxide Composites. *Environmental Science & Technology*, **47**, 6536-6544. http://dx.doi.org/10.1021/es400919t

[17] Li, X., Yu, J. and Jaroniec, M. (2016) Hierarchical Photocatalysts. *Chemical Society Reviews*, **45**, 2603-2636. http://dx.doi.org/10.1039/C5CS00838G

[18] Qi, K.Z., Yang, J.Q., Fu, J.Q., Wang, G.C., Zhu, L.J., Liu, G. and Zheng, W.J. (2013) Morphology-Controllable ZnO Rings: Ionic Liquid-Assisted Hydrothermal Synthesis, Growth Mechanism and Photoluminescence Properties. *CrystEngComm*, **15**, 6729-6735. http://dx.doi.org/10.1039/c3ce27007f

[19] Yu, W., Zhang, J. and Peng, T. (2016) New Insight into the Enhanced Photocatalytic Activity of N-, C- and S-Doped ZnO Photocatalysts. *Applied Catalysis B: Environmental*, **181**, 220-227. http://dx.doi.org/10.1016/j.apcatb.2015.07.031

[20] Rengaraj, S., Venkataraj, S., Tai, C., Kim, Y.H., Repo, E. and Sillanp, M. (2011) Self-Assembled Mesoporous Hierarchical-Like In_2S_3 Hollow Microspheres Composed of Nanofibers and Nanosheets and Their Photocatalytic Activity. *Langmuir*, **27**, 5534-5541. http://dx.doi.org/10.1021/la104780d

[21] Sajan, C.P., Wageh, S., Al-Ghamdi, A.A., Yu, J. and Cao, S. (2016) TiO_2 Nanosheets with Exposed {001} Facets for Photocatalytic Applications. *Nano Research*, **9**, 3-27. http://dx.doi.org/10.1007/s12274-015-0919-3

[22] Qi, K.Q., Wang, Y., Wang, R.D., Wu, D. and Li, G.D. (2016) Facile Synthesis of Homogeneous $CuInS_2$ Quantum Dots with Tunable Near-Infrared Emission. *Journal of Materials Chemistry C*, **4**, 1895-1899. http://dx.doi.org/10.1039/C5TC04232A

[23] Yu, W., Xu, D. and Peng, T. (2015) Enhanced Photocatalytic Activity of g-C_3N_4 for Selective CO_2 Reduction to CH_3OH via Facile Coupling of ZnO: A Direct Z-Scheme Mechanism. *Journal of Materials Chemistry A*, **3**, 19936-19947. http://dx.doi.org/10.1039/C5TA05503B

[24] Wetterskog, E., Tai, C.W., Grins, J., Bergström, L. and Salazar-Alvarez, G. (2013) Anomalous Magnetic Properties of Nanoparticles Arising from Defect Structures: Topotaxial Oxidation of $Fe_{1-x}O|Fe_{3-\delta}O_4$ Core|Shell Nanocubes to Single-Phase Particles. *ACS Nano*, **7**, 7132-7144. http://dx.doi.org/10.1021/nn402487q

[25] Low, J., Yu, J., Li, Q. and Cheng, B. (2014) Enhanced Visible-Light Photocatalytic Activity of Plasmonic Ag and Graphene Co-Modified Bi_2WO_6 Nanosheets. *Physical Chemistry Chemical Physics*, **16**, 1111-1120. http://dx.doi.org/10.1039/C3CP53820F

[26] Yang, Z.X., Zhang, P., Zhong, W., Deng, Y., Au, C.T. and Du, Y.W. (2012) Design, Growth, and Characterization of Morphology-Tunable $Cd_xZn_{1-x}S$ Nanostructures Generated by a One-Step Thermal Evaporation Process. *CrystEngComm*, **14**, 4298-4305. http://dx.doi.org/10.1039/c2ce25181g

[27] Li, Q., Meng, H., Zhou, P., Zheng, Y.Q., Wang, J., Yu, J.G. and Gong, J.R. (2013) $Zn_{1-x}Cd_xS$ Solid Solutions with Controlled Bandgap and Enhanced Visible-Light Photocatalytic H_2-Production Activity. *ACS Catalysis*, **3**, 882-889. http://dx.doi.org/10.1021/cs4000975

[28] Moon, G.D., Ko, S., Min, Y., Zeng, J., Xia, Y. and Jeong, U. (2011) Chemical Transformations of Nanostructured Materials. *Nano Today*, **6**, 186-203. http://dx.doi.org/10.1016/j.nantod.2011.02.006

[29] Li, W., Li, D., Zhang, W., Hu, Y., He, Y.H. and Fu, X. (2010) Microwave Synthesis of $Zn_xCd_{1-x}S$ Nanorods and Their Photocatalytic Activity under Visible Light. *The Journal of Physical Chemistry C*, **114**, 2154-2159. http://dx.doi.org/10.1021/jp9066247

[30] Han, D., Cao, J., Yang, S., Yang, J., Wang, B., Fan, L., Liu, Q., Wang, T. and Niu, H. (2014) Investigation of Composition Dependent Structural and Optical Properties of the $Zn_{1-x}Cd_xS$, Coaxial $Zn_{0.99-x}Cd_xCu_{0.01}S/ZnS$, $Zn_{0.99-x}Cd_xMn_{0.01}S$ Nanorods Generated by a One-Step Hydrothermal Process. *Dalton Transactions*, **43**, 11019-11026. http://dx.doi.org/10.1039/c4dt00667d

[31] Zhang, J., Shi, F., Chen, D., Gao, J., Huang, Z., Ding, X. and Tang, C. (2008) Self-Assembled

3-D Architectures of BiOBr as a Visible Light-Driven Photocatalyst. *Chemistry of Materials*, **20**, 2937-2941. http://dx.doi.org/10.1021/cm7031898

[32] Yang, F., Yan, N., Huang, S., Sun, Q., Zhang, L. and Yu, Y. (2012) Zn-Doped CdS Nanoarchitectures Prepared by Hydrothermal Synthesis: Mechanism for Enhanced Photocatalytic Activity and Stability under Visible Light. *The Journal of Physical Chemistry C*, **116**, 9078-9084. http://dx.doi.org/10.1021/jp300939q

[33] Hermann, J., Tahiri, H., Ait-Ichou, Y., Lassaletta, G., Gonzalez-Elipe, A. and Fernandez, A. (1997) Characterization and Photocatalytic Activity in Aqueous Medium of TiO_2 and Ag-TiO_2 Coatings on Quartz. *Applied Catalysis B: Environmental*, **13**, 219-228. http://dx.doi.org/10.1016/S0926-3373(96)00107-5

[34] Scalfani, A. and Hermann, J. (1998) Influence of Metallic Silver and of Platinum-Silver Bimetallic Deposits on the Photocatalytic Activity of Titania (Anatase and Rutile) in Organic and Aqueous Media. *Journal of Photochemistry and Photobiology A*, **113**, 181-188. http://dx.doi.org/10.1016/S1010-6030(97)00319-5

[35] Vamathevan, V., Amal, R., Beydoun, D., Low, G. and McEvoy, S. (2002) Photocatalytic Oxidation of Organics in Water Using Pure and Silver-Modified Titanium Dioxide Particles. *Journal of Photochemistry and Photobiology A*, **148**, 233-245. http://dx.doi.org/10.1016/S1010-6030(02)00049-7

[36] Li, X.Z. and Li, F.B. (2001) Study of Au/Au^{3+}-TiO_2 Photocatalysts toward Visible Photooxidation for Water and Wastewater Treatment. *Environmental Science & Technology*, **35**, 2381-2387. http://dx.doi.org/10.1021/es001752w

9

Indoor Air Environment—Hygienic Factors and Limits

Thomas Alsmo

Library Research, Olten, Switzerland
Email: administrator@libraryresearch.eu

Abstract

Complaints on the indoor environment of the residents in recent decades have become a common problem in the Swedish housing. The buildings themselves are said to be the cause of problems, and it is given a vague picture of both the exposure and the effect of the problems. The symptoms that residents and users state are often common in the population such as headache, fatigue, mucosal disorders and skin problems. It must be considered that the air that people routinely inhale contains impurities of various kinds, both in- and outdoors. An important source of contamination indoors is the microorganisms that are pathogenic, so called agents. Examples of infectious agents are viruses, fungi, bacteria and protozoa. The purpose of this project is to examine whether a physical measurement is possible to obtain for identifying a possible threshold level of air pollution in the indoor environment. In this study, carried out through physical measurements, the results show major deficiencies in the Swedish school environment. If we study the emissions in the important health-related size range of particles larger than 5.0 microns, before and after measures, the environmental benefits are clarified since over 90% of contaminants larger than 5.0 microns have been eliminated.

Keywords

Hygiene and Health, Indoor Air Quality, Indoor Humidity, Indoor Temperature

1. Introduction

Complaints on the indoor environment of the residents in recent decades have become a common problem in the Swedish housing. The buildings themselves are said to be the cause of problems, and it is given a vague picture of both the exposure and the effect of the problems. The symptoms that residents and users state are often common in the population such as headache, fatigue, mucosal disorders and skin problems. A fifth of the adult population in Sweden say they have one or more of these symptoms which they relate to the indoor environment in the home, school or work. There is a national authority on matters relating to safety, health and object-borne infection. These authorities

- are bringing forth directives and the requirements to be imposed on the indoor environment.
- compile the local supervisory work undertaken to create a comprehensive national picture of the problem area [1] [2] [3].

The Swedish authorities' annual monitoring of the area announced that complaints about the indoor environment are essentially about moisture and mold, ventilation and emissions and odor. The problems are so extensive that national enforcement projects are to come to terms with them. In the years 2014-2015, a project was focused on schools that are considered to be a sensitive environment, given the situation for children. The results show that many schools do not meet the law requirements to prevent problems with the indoor environment [4]. The basic legislation is the Environmental Code [5] and the Work Environment Act [6] which are the starting points for any oversight that occurs in health. They aim to promote sustainable development so that present and future generations can live in a healthy environment. Local supervisors should check that the laws are followed, provide information and advice and check that the activities are not involving a nuisance to human health [5] [6]. A basic problem is that it is not possible to determine what is causing health problems, that is, what kind of hygienic deficiency that may exist in the environment that cause people harm. Hygiene is a wide term which basically means cleaning the body or the human environment. What level of cleanliness that can be considered hygienic depends on what is the context. In a broader sense, the concept of activities aimed at removing any unwanted that is harmful or offensive, whether it physically exists in reality or whether it is more mentally linked. Hygiene can in this respect be a sensitive concept because it is not only about what is actually harmful, but also has to do with deeper psychological factors. However, the concern of this project is not the psychological factors, but only the physically real emissions that spread illness in buildings. The air that people normally breathe regularly contains of pollution of various kinds, both indoors and outdoors. Air pollution is present in two forms, gases and particles. Content and distribution are factors that affect air quality. An important source of contamination indoors is the microorganisms that are pathogenic and examples are viruses, fungi, bacteria and protozoa. These are often spread through airborne particles, preferably those larger than 5.0 microns. That means these particles are an important factor in the air since they are carriers of these microorganisms. Moreover, considered the situation for those suffering from allergy and asthma, these particles are also spreaders of allergens [7]. Particles in the air also affect the well-being of people through the respiratory system, cardiovascular system and studies show the link with lung cancer. It should be stressed that the impact on health is proportional to the concentration of particles. Even brief exposure to high concentrations of PM10 (PM10 = particles less than 10 microns) can lead to serious effects [8]. Overall, it is important that the amount of particles can be kept as low as possible. The base of the indoor air is the air surrounding a building. Regarding particles they are a normally occurring air pollution and outdoors are several different particle sources, including the burning of fossil fuels and biofuels, road and vehicle wear and tear, road sand, forest fires, sea, desert sand and other soil materials. Emission sources vary between countries and especially the size determines how far a particle can be transported. Particles less than 10 micrometers (PM10.0) can be divided into three different size ranges, *coarse*, *fine* and *ultrafine*.

1) Coarse particles consist of sizes between about PM10 (less than 10.0 microns) and PM1 (less than 1.0 microns) and consist mainly of particles formed by including wear and tear of the roadway, brakes, tires, and sand on roads and streets.
2) Fine particulate matter consists of sizes between about PM1 and PM0.1 (less than 0.1 microns) and in Sweden these particles originate mainly from emissions in other countries. Whence arrival is determined by the wind because the fine particles can stay in the atmosphere a number of days.
3) Ultra-fine particles are particles smaller than about PM0.1. These come from local vehicles, wood burning and energy facilities. The ultra-fine particles are numerically the majority of the total PM10 level [8].

As for the gaseous pollutants, it has not demonstrated any dose-response relationship, despite intensive research in recent decades in the environments that this project concerns [9]-[26]. With a dose-response relationship is meant the concentration of a substance *i.e.* pollution required to inconvenience or disease to occur. A fundamental problem for the indoor environment research is the lack of validated methods and that there are no validated limits for the air environment indoors for residential. There is a great need to be able to physically measure the indoor air environment with a view to putting a limit on the emission level indoors in homes, schools and offices.

2. Measurements

The purpose of this project is to investigate whether through physical measurements we can identify a possible threshold level of air pollution in the indoor environment. The project includes physical measurements in 45 schools, school areas where students normally stay. The school buildings constitute a representative sample of the Swedish conditions, both in number and in geographic distribution across the country, to obtain an idea about the hygienic status of the indoor environment in Swedish schools. It is necessary to create a suitable level sensor or appraiser for indoor air that clarifies whether deficiencies exist or not. For this purpose, it is examined whether the outdoor air can be used as a parameter to be able to make comparisons between different indoor environments. Within the framework of this project, measurements are performed in outdoor air at the 45 sites where school buildings are located. The average of these measurements are used and then evaluated if this might be the appropriate method to use as a parameter in the evaluation of indoor air. Emission measurements are performed using the TSI instrument laser sampler model 8220. The instrument measures six compartment sizes particles, 0.3, 0.5, 1.0, 3.0, 5.0 and 10.0 microns. At each test site an instrument is programmed for a measurement time of one minute per measurement. A series of measurements comprises five individual measurements, which means that approximately 15 liters of air is treated. For the five individual values are calculated an average for all test sites for each school and used as a comparison value in the assessment of the various school environments. The results of the measurements in this project are reported in the number of particles per cubic meter of air. The measurement is complemented by visual inspection of the school premises and the objective is to visually identify sources of hygiene deficiencies. If it appears from the measurements and the visual inspection that there is need for improvements to reduce the amount of pollutants in the air, a few representative buildings are selected from the 45 units that are fixed with the aim of reducing the amount of pollutants in the air. All the results from the measurement in the school environment is based on an activity that is comparable to the activity normally pursued in the school environment, but with an important difference and that is the number of people present. When measurements are conducted only a few people are involved, unlike during normal activities when far more people are staying at the premises. It is important to know that each person affects the environment. All cells of a human being, except for neurons, are continually renewed. Old skin is continuously replaced by newly formed skin.

As a rule a normal person exchanges an entire outer layer of skin after four days. The following number of particles larger than 0.5 microns, the body gets rid of during different movements [20].

- Sedentary: 100 000 particles per minute.
- Seated, rotating arm movements: 500,000 particles per minute.
- Stand up, sit down: 2.5 million particles per minute.
- Climbing stairs, running: 10,000,000 particles per minute.

This means that the emission supplement that normally occurs when the activity is in progress will not be included in the measurements in this study.

3. Results

The base of the air environment indoors is the air surrounding the building and as shown in **Table 1** and in the graph in **Figure 1** the emission level in the outdoor environment is high in the smallest size range. As shown in **Table 1** the levels vary widely, preferably in the smaller ranges to subside in the larger ones. To eliminate the various instantaneous changes in the outside air from one point to another, the average value of all the 45 locations is used. The reason for this is that it evens out the level and variety of extreme values can be excluded that would otherwise pose a risk in a direct comparison with the outdoor environment of the respective test site. The compilation of the average value of the buildings studied in this project is presented in **Table 2** and with the graphs in **Figure 2** where the emission amount indoors and outdoors is compared. The graphs in **Figure 2** show higher emission level outdoors in the smallest size range, to the greater ranges diverge to the detriment of the indoor environment. The reason

Table 1. Particulate emissions from outdoor air, each fraction larger than 0.3 - 10.0 microns (number of particles per cubic meter), the average value of a series of measurements at 45 locations in Sweden.

Unit	>0.3	>0.5	>1.0	>3.0	>5.0	>10.0
1	27,725,000	1,749,000	718,000	15,000	5000	1000
2	50,277,000	4,020,000	174,000	29,000	12,000	4000
3	20,998,000	2,079,000	256,000	84,000	27,000	3000
4	53,491,000	3,882,000	151,000	27,000	10,000	2000
5	22,644,000	2,015,000	229,000	72,000	25,000	4000
6	23,572,000	1,703,000	191,000	62,000	27,000	4000
7	34,004,000	3,912,000	818,000	384,000	174,000	29,000
8	17,828,000	4,502,000	1,520,000	669,000	56,000	9000
9	21,097,000	3,809,000	485,000	112,000	42,000	9000
10	28,424,000	2,216,000	172,000	53,000	16,000	3000
11	4,104,000	461,000	83,000	30,000	10,000	2000
12	6,800,000	1,703,000	314,000	17,000	3000	1000
13	12,111,000	4,995,000	1,489,000	165,000	30,000	1000
14	38,473,000	3,523,000	287,000	61,000	21,000	5000
15	6,194,000	708,000	78,000	22,000	8000	0
16	101,243,000	17,675,000	900,000	88,000	40,000	20,000
17	8,037,000	663,000	145,000	52,000	20,000	4000
18	19,694,000	12,980,000	3,010,000	658,000	73,000	1000
19	4,845,000	1,775,000	748,000	98,000	16,000	2000
20	2,905,000	739,000	380,000	188,000	95,000	25,000
21	6,139,000	520,000	104,000	4000	1000	0
22	24,007,000	1,236,000	152,000	40,000	14,000	4000
23	6,783,000	1,665,000	801,000	204,000	71,000	13,000
24	12,283,585	1,418,047	479,385	59,951	19,362	5580
25	1,600,300	511,057	209,921	53,823	22,131	4895
26	2,867,000	138,806	69,046	23,627	9293	1844
27	2,371,000	1,380,000	256,609	84,874	38,410	8364
28	6,992,000	1,902,000	609,000	78,000	23,000	4000
29	7,835,000	3,377,000	1,213,000	193,000	52,000	14,000
30	198,900,000	17,796,000	2,115,000	188,000	31,000	2000
31	5,646,000	1,814,000	641,000	78,000	21,000	2000

Continued

32	2,882,000	982,000	651,000	82,000	15,000	1000
33	1,631,000	224,000	138,000	64,000	12,000	0
34	6,254,000	999,000	226,000	37,000	12,000	2000
35	54,620,000	6,429,000	654,000	136,000	65,000	11,000
36	18,402,000	2,285,000	341,000	48,000	15,000	3000
37	17,432,000	2,001,000	719,000	150,000	6000	2000
38	39,711,000	3,030,000	185,000	29,000	15,000	4000
39	29,982,000	2,790,000	437,000	98,000	32,000	3000
40	54,912,000	3,288,000	261,000	17,000	9000	2000
41	22,614,000	3,901,000	249,000	172,000	24,000	4000
42	25,712,000	1,356,000	201,000	159,000	29,000	4000
43	34,431,000	4,023,000	799,000	285,000	89,000	29,000
44	7,801,000	2,073,000	401,000	16,000	3000	1000
45	26,430,000	4,011,000	419,000	128,000	34,000	3000
Average outdoors	24,949,000	3,206,000	544,000	118,000	30,000	6000

Table 2. Particulate emissions in 45 schools, each fraction larger than 0.3 - 10.0 microns (number of particles per cubic meter), the average value of a measurement series.

Unit	>0.3 μm	>0.5 μm	>1.0 μm	>3.0 μm	>5.0 μm	>10.0 μm
1	8,138,000	5,014,000	4,374,000	3,856,000	3,286,000	2,297,000
2	36,798,000	30,056,000	28,375,000	25,383,000	20,752,000	12,480,000
3	13,808,000	8,792,000	7,629,000	6,604,000	5,346,000	3,145,000
4	67,085,000	57,661,000	52,531,000	44,348,000	33,189,000	15,219,000
5	21,104,000	11,210,000	9,491,000	8,084,000	6 300 000	3,386,000
6	9,979,000	1,942,000	1,026,000	711,000	515,000	315,000
7	21,366,000	16,607,000	15,140,000	13,252,000	10,723,000	6,436,000
8	23,509,000	17,340,000	16,303,000	15,132,000	12,930,000	7,681,000
9	13,259,000	4,301,000	2,429,000	1,644,000	1,195,000	653,000
10	15,909,000	11,935,000	8,580,000	7,435,000	6,283,000	3,520,000
11	9,774,000	6,820,000	5,655,000	4,805,000	4,043,000	2,512,000
12	18,834,000	14,396,000	11,961,000	10,211,000	7,792,000	4,508,000
13	10,996,000	6,252,000	3,599,000	3,000,000	2,703,000	1,898,000
14	4,360,000	3,065,000	2,619,000	2,356,000	1,983,000	1,194,000
15	16,015,000	10,613,000	7,335,000	5,974,000	4,691,000	2,550,000
16	16,489,000	6,770,000	5,168,000	3,751,000	2,707,000	1,513,000

Continued

17	3,959,000	2,001,000	1,919,000	1,815,000	1,674,000	1,208,000
18	3,922,000	1,221,000	935,000	805,000	698,000	523,000
19	5,268,000	4,899,000	4,564,000	3,469,000	2,388,000	1,014,000
20	5,811,000	2,859,000	2,246,000	1,695,000	1,216,000	676,000
21	4,755,000	2,791,000	1,945,000	1,333,000	898,000	444,000
22	12,282,000	8,158, 00	6,599,000	5,661,000	4,803,000	3,037,000
23	51,320,000	16 484 000	11 664,000	9,884,000	8,111,000	4,948,000
24	52,420,000	11 166 000	5,787,000	4,749,000	3,730,000	2,085,000
25	51,017,000	12,511,000	7,355,000	5,689,000	4,108,000	1,987,000
26	34,325,000	22,520,000	20,997,000	19,170,000	16,338,000	10,566,000
27	9,328,000	4,418,000	3,864,000	3,466,000	2,957,000	2,026,000
28	9,500,000	7,315,000	5,849,000	4,182,000	3,074,000	2,007,000
29	25,652,000	22,650,000	20,095,000	16,494,000	12,671,000	6,817,000
30	5,450,000	4,077,000	3,678,000	3,211,000	2,532,000	1,498,000
31	14,511,000	3,399,000	2 147 000	1 449 000	1,016,000	556,000
32	9,710,000	7,308,000	5,980,000	3,988,000	2,587,000	1,329,000
33	10,927,000	8,564,000	7,575,000	6,738,000	5,763,000	4,315,000
34	45,557,000	42,070,000	37,797,000	32,378,000	23,330,000	11,659,000
35	7,116,000	5,699,000	4,727,000	3,582,000	2,527,000	1,421,000
36	2,838,000	2,642,000	2,410,000	1,895,000	1,373,000	767,000
37	14,002,000	11,062,000	9,020,000	7,637,000	6,216,000	3,617,000
38	16,423,000	11,438,000	10,308,000	4,126,000	2,247,000	1,065,000
39	17,604,000	14,334,000	12,278,000	10,282,000	6,127,000	2,886,000
40	19,896,000	16,650,000	14,912,000	12,726,000	9,731,000	6,166,000
41	18,551,000	12,509,000	10,985,000	9,253,000	6,978,000	4,070,000
42	27,714,000	18,386,000	10,751,000	6,200,000	3,254,000	1,089,000
43	3,719,000	3,337,000	3,013,000	2,262,000	1,568,000	730,000
44	53,360,000	49,053,000	46,568,000	40,470,000	30,792,000	15,727,000
45	30,022,000	11,247,000	8,414,000	6,797,000	5,249,000	2,700,000
Average indoors	19,431,000	12,301,000	10,369,000	8,623,000	6,631,000	3,694,000

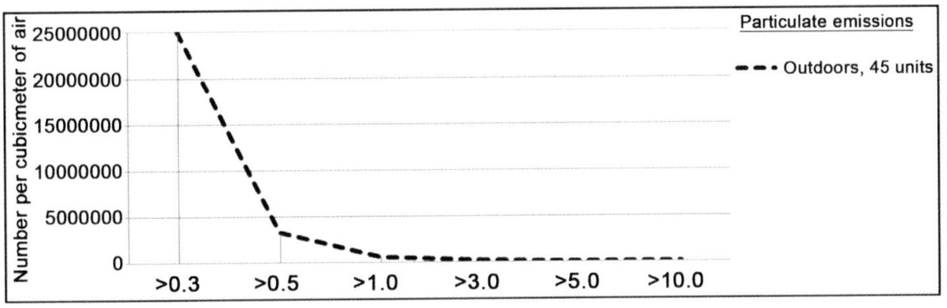

Figure 1. Emission level of outdoors, open interval larger than 0.3 - 10 microns, number of particles per cubic meter, average value from 45 geographically distributed sites in Sweden.

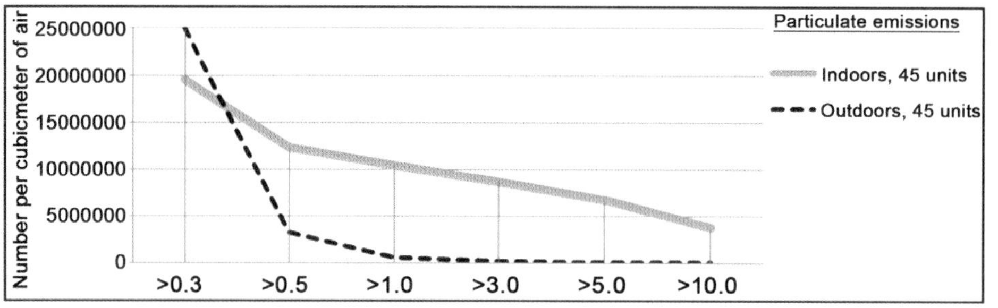

Figure 2. Emission level, open interval larger than 0.3 - 10 microns, number of particles per cubic meter, average value of outdoor air around the building and indoor air in 45 schools.

for these high emission levels indoors in the larger interval is not due to outside air, but depends on the emission sources available in the environment. An important source of contamination indoors is the microorganisms that are pathogenic, called agents, and examples of infectious agents, viruses, fungi, bacteria and protozoa. These are often spread through airborne particles, those larger than 5.0 microns. That means these particles are an important factor in the air as they are carriers of these microorganisms. Moreover, considered the situation for people suffering from allergy and asthma these particles are also spreaders of allergens [7] [20] [27]-[47]. The National Public Health Institute in Sweden showed already in 1995 that the number of particles smaller than 1.0 microns will constitute the entire 99.9% of all particles in the air [28].

In conjunction with the visual inspections of the buildings it revealed that there are a variety of materials in schools that are not used, but that it is both generates emissions and causes the environment, from a hygienic point of view, becoming overloaded and cannot be maintained to the extent required. Four of the above school environments were selected as representative units regarding emission load and where measures were taken to correct the identified deficiencies. The graphs in **Figure 3** show the emission levels of outdoor air for all 45 units and for the four selected units separately, indicating that a similar load level prevails in the air environment as the average for all 45 units. The measures taken in the 4 units consisted of cleaning up and remove the material that was not needed. Then, there are better opportunities for appropriate handling and storage of the material remained. Through these measures, opportunities are created from a hygiene perspective maintains the premises to the extent necessary, and then new measurements can be performed. Significant improvements have been achieved, results can be seen in **Table 3** and graphs in **Figure 4**, which show that the emission level indoors after measures follows the outdoor level in a satisfactory manner. The study has been conducted with physical measurements have reported significant deficiencies in the Swedish school environment. If we study the important size range larger than 5.0 microns, before and after measures, the environmental benefits is made clear since over 90% of contaminants larger than 5.0 microns have been eliminated. **Figure 5** shows all particles larger than 5.0 microns, before and after measures, of the four selected units. The result is striking, emission load of negative health topics have been reduced by over 90%. **Figure 6** reports using a logarithmic scale that allows direct comparison of each

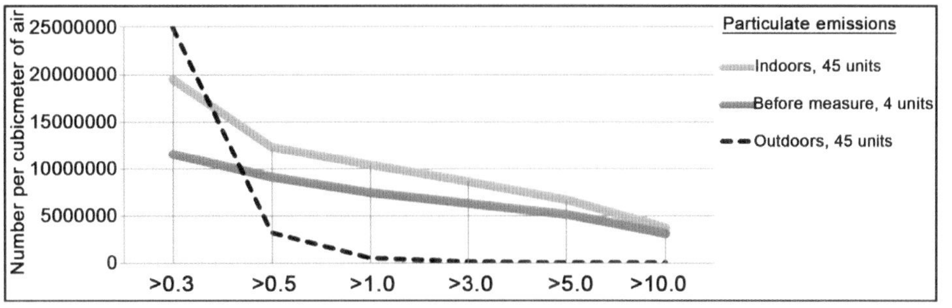

Figure 3. Emission level, open interval larger than 0.3 - 10 microns, number of particles per cubic meter, average value of outdoor air around the building and indoors in 45 schools, and from these four selected schools that are recognized separately.

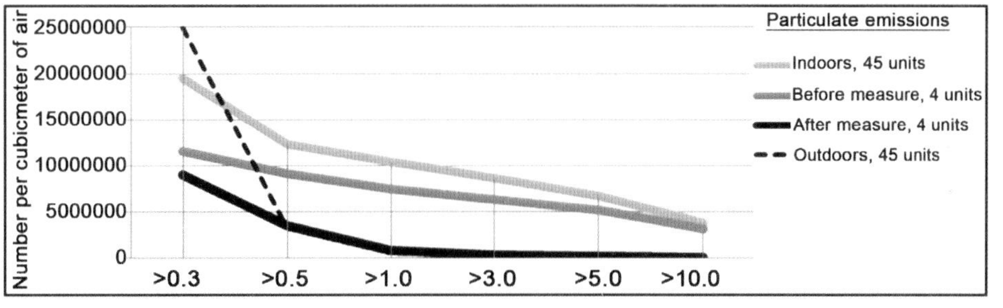

Figure 4. Emission level, open interval larger than 0.3 - 10 microns, number of particles per cubic meter, average value of outdoor air around the building and indoors in 45 schools. From these 45 buildings 4 selected schools that are recognized separately in the diagram before and after measure.

Table 3. Particulate emissions, each fraction larger than 0.3 - 10.0 microns (number of particles per cubic meter), the average value for four schools, before and after measure.

	>0.3	>0.5	>1.0	>3.0	>5.0	>10.0
Before						
Unit 10	15,909,000	11,935,000	8,580,000	7,435,000	6,283,000	3,520,000
Unit 19	5,268,000	4,899,000	4,564,000	3,469,000	2,388,000	1,014,000
Unit 33	10,927,000	8,564,000	7,575,000	6,738,000	5,763,000	4,315,000
Unit 37	14,002,000	11,062,000	9,020,000	7,637,000	6,216,000	3,617,000
Average	11,526,500	9,115,000	7,434,750	6,319,750	5,162,500	3,116,500
After						
Unit 10	18,271,000	10,211,000	2,587,000	517,000	197,000	48,000
Unit 19	4,473,000	1,434,000	109,000	64,000	34,000	16,000
Unit 33	5,998,000	843,000	202,000	101,000	68,000	41,000
Unit 37	7,141,000	1,686,000	396,000	612,000	459,000	272,000
Average	8,970,750	3,543,500	823,500	227,333	189,500	94,250

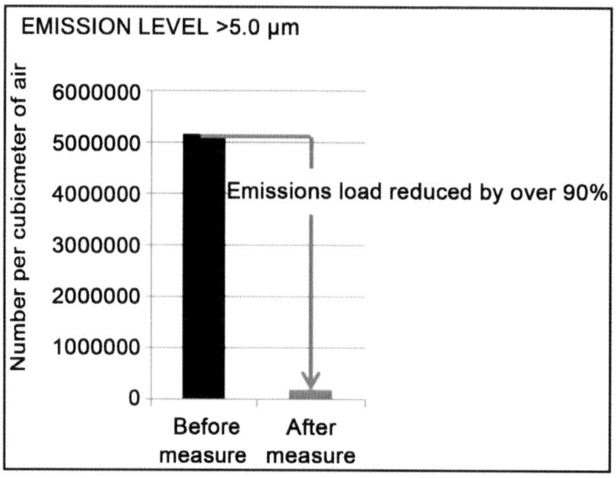

Figure 5. Particulate emissions in the open interval larger than 5.0 microns, before and after measure.

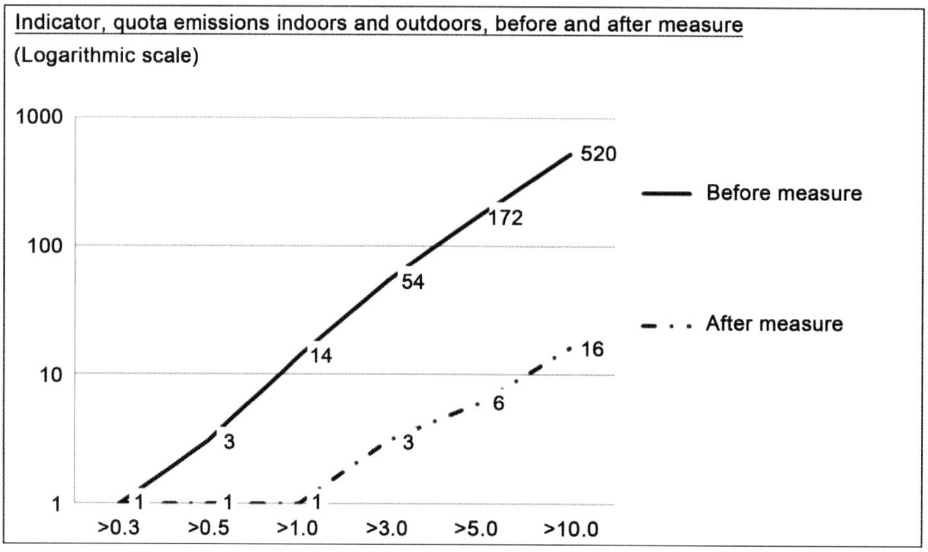

Figure 6. Quota, particulate air pollution, possible level for each fraction larger than 0.3 - 10.0 microns. Results for indoor- and outdoor air from four schools, before and after measure.

fraction and clarifies the big health gain achieved. The results of this study show that the curve for action in **Figure 6** is a perfectly level sensor for how many emissions should be tolerated in the Swedish residential and school environment. These results are also shown in **Table 4**.

4. Conclusions

The results presented in this project show that outdoor air can be advantageously helpful in the evaluation of the indoor environment. By using a plurality of values from the outdoor environment, a neutral value is created, a constant, which eliminates the instantaneous fluctuations that occur in the outdoor air from time to another. We use the average of all 45 sites outliers removed and instead use this constant to the respective

Table 4. Quota, particulate air pollution, possible level for each fraction larger than 0.3 - 10.0 microns. Results for indoor- and outdoor air from four schools, before and after measure.

	>0.3	>0.5	>1.0	>3.0	>5.0	>10.0 μm
Before						
Unit 10	0	4	16	63	209	587
Unit 19	0	1	8	29	80	169
Unit 33	0	3	14	57	192	719
Unit 37	0	3	17	65	207	603
After						
Unit 10	0	3	5	4	7	8
Unit 19	0	0	0	0	1	3
Unit 33	0	0	0	1	2	7
Unit 37	0	0	0	5	15	45

measured level of a building, the calculation done by the simple formula, "amount of emission indoors divided by constant". The four selected units that are used to evaluate this process recognize that the results are clarified by the difference between the two graphs in **Figure 6** drastically increasing in the larger particle size fractions. The level values for emissions larger than 5.0 microns are 172 before measures and 6 after measures and emission levels for larger than 10.0 microns are 520 before measures and 16 after measures. As the results of this study show the air environment in the Swedish school buildings is congested and unhealthy particulate emissions are very high compared to what is possible to achieve. It indicates that there are some deficiencies in the Swedish system. **Figure 5** shows that the emission amount can be significantly improved in terms of health in the sensitive range larger than 5.0 microns. The emission load has been reduced by over 90% and this with cost-effective measures. An important question is how the current situation of high emission levels can proceed. Results show that this relationship was identified already in the 1990s, with no reaction from the Swedish authorities, who maintained the building focused direction, without further examination [48]. In direct inquiries to the relevant authorities on two occasions, in 2007 and 2014, what they do to remedy the grievance that this problem was neglected. Instead, they refer to others, in this context, non-validated parameters and measurement methods, with their view that measures must be made in the building construction [49] [50] [51]. A basic question is who owns the problem, the business or the property owner. According to the authorities, property owners in Sweden has a so-called reverse burden of proof, *i.e.* when the diffuse health problems related to indoor environment occurs, the property owner shall report to the Authority that they, as the authorities put it, "have a functioning self-control" [52]. The situation is often difficult to interpret when problems occur. What does the authority require from the property owner? This is by no means clear. However property owners are, for unclear reasons, forced to extensive technical measures by rebuilding and strengthening ventilation with mechanical systems. These are measures that often require large financial resources. The results achieved are difficult to identify and problems usually reappear after some time with new requirements on the property owner [53] [54] [55] [56]. It is very important for the development of the indoor environment in Swedish buildings, to critically examine the methods used today. Can there be deficiencies in the working models and analysis procedures used? Are important parameters overlooked that affect the indoor environment and human health, but not related to deficiencies in building design? The working model must not be too tight, which means that critical parameters are not included, which misleads development. There are fundamental deficiencies in the working methods that the authorities now require. In this context, not performing necessary technical measures will ultimately affect those staying in the building [51] [57] [58] [59] [60].

- It is important to state what affects the environment in buildings, *i.e.* which is the emission sources.
- The hypothesis should not only focus on people, with an emphasis on the health problems found in an inadequate building construction.
- What are the conditions that the working model to work as crucial parameters omitted?

The model must of course be complete and correctly account for all the sources of the problems. In this context, it means to take into account the total human environment, the activities in the building and the building itself. Apart from the large waste of resources that occurs in connection with meaningless technical measures, it becomes ultimately the occupants of the premises that suffer when the real weaknesses are not taken into account since the actual emission sources are not taken into account.

References

[1] Merritt, A.S. and Emenius, G. (2013) Environmental Health Report (Miljöhälsorapport). Institute of Environmental Medicine, Karolinska Institutet, Stockholm, Sweden. (In Swedish)

[2] The Public Health Agency of Sweden. (In Swedish)
https://www.folkhalsomyndigheten.se/livsvillkor-levnadsvanor/miljohalsa-och-halsoskydd/inomhusmiljo-allmanna-lokaler-och-platser/
https://www.folkhalsomyndigheten.se/livsvillkor-levnadsvanor/miljohalsa-och-halsoskydd/inomhusmiljo-allmanna-lokaler-och-platser/tillsynsvagledning-miljobalken/

[3] Swedish Work Environment Authority. Sick-Buildings.
https://www.av.se/inomhusmiljo/inomhusmiljo-och-halsobesvar/

[4] The Environmental Goals. Annual Follow-Up of Sweden's Environmental Quality Objectives and Milestones 2016, Report 6707 Sides 267 - 268, Mars 2016, The Swedish Environmental Protection Agency.

[5] Environmental Code (Miljöbalken), Swedish Code of Statutes 1998:808 (Svensk författningssamling 1998:808), Ministry of the Environment and Energy (Miljö-och energidepartementet), Government Offices (Regeringskansliet/Lagrummet.se). (In Swedish)
http://www.riksdagen.se/sv/dokument-lagar/dokument/svensk-forfattningssamling/miljobalk-1998808_sfs-1998-808

[6] The Work Environment Act (Arbetsmiljölagen), Swedish Code of Statutes 1977:1160 (Svensk

författningssamling 1977:1160), Ministry of Labour (Arbetsmarknadsdeartimentet), (Regeringskansliet/Lagrummet.se). (In Swedish)
https://www.riksdagen.se/sv/dokument-lagar/dokument/svensk-forfattningssamling/arbetsmiljolag-19771160_sfs-1977-1160

[7] Ljungqvist, B., Nydahl, R. and Reinmüller, R. (1989) Measurement in Rooms with Increased Purity Requirements (Mätteknik i rum med förhöjda renhetskrav). Royal Institute of Technology in Stockholm och Kabi Stockholm. (In Swedish)

[8] Lundberg, I. and Johansson, J. (2002) The Amount of Harmful Particles in the Air Must Be Reduced (Mängden skadliga partiklar i luften måste minska). County Board in Stockholm (Länsstyrelsen), Sweden.

[9] Wålinder, R. and Friis, L. (1999) Medical Investigation of Indoor Related Disorders 5/99 (Medicinsk utredning av inomhusrelaterade besvär). Occupational and Environmental Medicine, Uppsala, Sweden. (In Swedish)

[10] Brauer, C. (2005) The Sick Building Syndrome Revisited. Department of Occupational Medicine, Copenhagen University, Denmark, 2005

[11] Thörn, Å. (1999) The Emergence and Preservation of Sick Building Syndrome. Department of Public Health Sciences, Division of Social Medicine, Karolinska Institute.

[12] Stridh, G. (2003) Are We Measuring the Chemistry for Floor Coverings and How We Interpret in This Case, Measurement Results (Skall vi mäta kemin under golvbeläggningar och hur tolkar vi I så fall mätresultat)? Inomhusklimat 203, sidor 102 - 108, A Wellness and Healthcare Center, Örebro University Hospital, Sweden. (In Swedish)

[13] Eriksson, B., Snaar, J. and Wärn, B. (1994) Environmental and Quality of Building Repairs (Miljö-och kvalitetsmärkning av byggreparationer). Sides 22 - 23, Folksams Research Foundation, Stockholm, Sweden. (In Swedish)

[14] Alsmo, T. (1994) A Study with Questionnaires and Particle Measurements (En studie med enkäter och partikelmätningar). Division of Construction Management, Royal Institute of Technology, Stockholm, Sweden. (In Swedish)

[15] Alsmo, T. and Holmberg, S. (2007) Sick Buildings or Not: Indoor Air Quality and Health Problems in Schools. *Indoor and Built Environment*, **16**, 548-555.

[16] Alsmo, T. and Holmberg, S. (2010) A Study of Sources of Airborne Pollutants and Poor Hygiene in Schools. *Indoor and Built Environment*, **19**, 298-304.

[17] Alsmo, T. and Alsmo, C. (2013) A Study of Hygiene in Swedish Schools and Pre-Schools-Sources of Air Pollution. *Journal of Environmental Protection*, **4**, 1349-1359.

[18] Alsmo, T. and Alsmo, C. (2014) Ventilation and Relative Humidity in Swedish Buildings. *Journal of Environmental Protection*, **5**, 1022-1036.

[19] Alsmo, T. and Alsmo, C. (2016) A Comparison of Relative Humidity between Two Swedish Buildings with Different Ventilation Solutions. *Journal of Environmental Protection*, **7**, 855-873.

[20] Ramstorp, M. (1999) Purity Technique and Clean Rooms—An Introduction (Renhetsteknik och rena rum—En introduction). Sides 17 - 18, Bioteknisk Processdesign AB, Malmö, Sweden. (In Swedish)

[21] Ekstand-Tobin, A. (1993) Correlation between Asthma and Indoor Environments (Samband mellan astma och inomhusmiljö). Borås, Building Research and SP Technical Research Institute of Sweden. (In Swedish)

[22] OSHA (1989) Carbon Dioxide, Comments from the January 19, 1989, Final Rule on Air Contaminants Project Extracted from 54FR2332 et. seq. This Rule Was Remanded by the U.S. Circuit Court of Appeals and the Limits Are Not Currently in Force.

[23] Berglund, B. and Johansson, B. (1996) Health Effects of Volatile Organic Compounds in Indoor Air. Stockholm's Universities and Karolinska Institute, Stockholm.

[24] Bornehag, C.G. and Stridh, G. (1995) Volatile Organic Compounds (VOC) in Indoor Air in the Swedish Housing (Flyktiga organiska ämnen (VOC) i inomhusluften i det svenska bostadsbeståndet). Karlstad University, Örebro University Hospital, Sweden. (In Swedish)

[25] Bornehag, C.G. (2000) Damp Buildings and Health (Fukt i byggnader och hälsa). Building Research, Stockholm, Sweden. (In Swedish)

[26] Bornehag, C.G., Sundell, J., Bonini, S., Custovic, A., Malmberg, P., Skerfing, S., Sigsgaard, T. and Verhoeff, A. (2004) Dampness in Buildings as a Risk Factor for Health Effects, EUROEXPO: A Multidisciplinary Review of the Literature (1998-2000) on Dampness and Mite Exposure in Buildings and Health Effects. *Indoor Air*, **14**, 243-257. http://dx.doi.org/10.1111/j.1600-0668.2004.00240.x

[27] Månsson, L. (1991) High-Tech Invisible Enemy (Högteknologins osynliga fiende). Corona AB, Malmö, Sweden. (In Swedish)

[28] Sundell, J. and Kjellman, M. (1995) The Air We Breathe Indoors (Luften vi andas inomhus). Public Health Institute (Folkhälsoinstitutet), Sweden. (In Swedish)

[29] Couch, R.B. (1981) Viruses and Indoor Air Pollution. *Bulletin of the New York Academy of Medicine*, **57**, 907-921.

[30] Smith, E.B. (1983) Atmospheric Factors Affecting Transmission of Infections. *The Practitioner*, **227**, 1667-1677.

[31] Hatch, M.T. and Wolochow, H. (1969) Bacterial Survival: Consequences of the Airborne State. In: Dimmick, R.L. and Akers, A.B., Eds., *An Introduction to Experimental Aero Biology*, John Wiley and Sons, New York.

[32] Anderson, J.D., Dark, E.A. and Pbto, S. (1968) The Effect of Aerosolization upon Survival and Potassium Retention by Various Bacteria. *Microbiology*, **52**, 99-105. http://dx.doi.org/10.1099/00221287-52-1-99

[33] Rosebury, T. (1947) Experimental Airborne Infection. Williams and Wilkins, Baltimore.

[34] Langmuir, A.D. (1980) Changing Concepts of Airborne Infection of Acute Contagious Diseases: A Reconsideration of Classic Epidemiologic Theories. *Annals of the New York Academy of Sciences*, **353**, 35-44. http://dx.doi.org/10.1111/j.1749-6632.1980.tb18903.x

[35] Moffet, H.L. (1980) Clinical Microbiology. J. B. Lippincott, Philadelphia.

[36] Gwaltney, J.M. (1980) Epidemiology of the Common Cold. *Annals of the New York Academy of Sciences*, **353**, 54-60. http://dx.doi.org/10.1111/j.1749-6632.1980.tb18905.x

[37] Harper, G.J. (1961) Airborne Micro-Organisms: Survival Tests with Four Viruses. *Epidemiology and Infection*, **59**, 479-486. http://dx.doi.org/10.1017/S0022172400039176

[38] Schulman, J.L. and Kilbourne, E.D. (1962) Airborne Transmission of Influenza Virus Infection in Mice. *Nature*, **182**, 1129-1130. http://dx.doi.org/10.1038/1951129a0

[39] Korsgaard, J. (1982) Preventive Measures in House-Dust Allergy. *American Review of Respiratory Disease*, **125**, 80-84.

[40] Arlian, L.G., Bernstein, I.L. and Gallagher, J.S. (1982) The Prevalence of House Dust Mites, *Dermatophagoides* spp, and Associated Environmental Conditions in Homes in Ohio. *Journal of Allergy and Clinical Immunology*, **69**, 527-532. http://dx.doi.org/10.1016/0091-6749(82)90178-6

[41] Murray, A.B. and Zuk, P. (1979) The Seasonal Variation in a Population of House Dust Mites in a North American City. *Journal of Allergy and Clinical Immunology*, **64**, 266-269. http://dx.doi.org/10.1016/0091-6749(79)90142-8

[42] Arlan, L.G., Brandt, R.L. and Bernstein, R. (1978) Occurrence of House Dust Mites, *Der-*

matophagoides spp. during the Heating Season. *Journal of Medical Entomology*, **15**, 35-42. http://dx.doi.org/10.1093/jmedent/15.1.35

[43] English, P.T. (1980) Medical Mycology. Edward Arnold, London.

[44] NAS (National Academy of Sciences) (1981) Indoor Pollutants. National Academy Press, Washington DC.

[45] IEC Beak Consultants Ltd. (1983) Indoor Air Quality, Cambridge Sealed Homes: A Report for Ontario Ministry of Municipal Affairs and Housing. IEC Beak, Mississauga, Ontario, October.

[46] Sheppard, D., Wong, W.S. and Uehora, C.F. (1981) Lower Threshold and Greater Bronchomotor Responsiveness of Asthmatic Subjects to Sulfur Dioxide. *American Review of Respiratory Disease*, **122**, 873-878.

[47] Zeterberg, J.M. (1973) A Review of Respiratory Virology and the Spread of Virulent and Possibly Antigenic Viruses via Air Conditioning Systems. *Annals of Allergy*, **31**, 228-234.

[48] Becker, M. (1996) The National Board of Health and Welfare (Socialstyrelsen). Indoor Environmental Studies at a nursery in Vårgårda Municipality (Inomhusmiljöundersökningar vid ett daghem I Vårgårda kommun).

[49] Lundh, A.L. (2007) Swedish Work Environment Authority. Solna, Sweden, Diarienummer: CTB 2007/35764.

[50] Haux, F. (2014) The Public Health Agency of Sweden (Folkhälsomyndigheten). Correspondence 2014-02-11.

[51] SWESIAQ, Swedish Chapter of International Society of Indoor Air Quality and Climate. http://www.swesiaq.se/
http://www.swesiaq.se/swesiaq-modellen.aspx

[52] The Public Health Agency of Sweden.
https://www.folkhalsomyndigheten.se/publicerat-material/publikationer/Fastighetsagarnas-egenkontroll/

[53] Henåns School, Orust Municipality, Sweden (2012). www.orust.se

[54] Ljungbyholms Preschool, Kalmar Municipality (2011). www.kalmar.se

[55] Tallparksgårdens Retirement Home, Östhammars Municipality (2012). www.osthammar.se

[56] City Hall, Falu Municipality (2015). www.falun.se

[57] Blom, N. (2014) General Advice on Ventilation. The Public Health Agency's, FoHMFS 2014:18. The Public Health Agency of Statutes, Solna, Sweden.
https://www.folkhalsomyndigheten.se/publicerat-material/foreskrifter-och-allmanna-rad/fohmfs-201418/

[58] Blom, N. (2014) The Public Health Agency of Public Moisture and Microorganisms, FoHMFS 2014:14. The Public Health Agency of Statutes, Solna, Sweden.
https://www.folkhalsomyndigheten.se/publicerat-material/foreskrifter-och-allmanna-rad/fohmfs-201414

[59] Swedish Work Environment Authority, Workplace design, Provisions AFS 2009:2.
https://www.av.se/globalassets/filer/publikationer/foreskrifter/arbetsplatsens-utformning-foreskrifter-afs2009-2.pdf

[60] Boverket, National Board of Housing, Building and Planning (2016).
http://www.boverket.se/sv/boende/halsa--inomhusmiljo-i-ditt-boende/luft-och-ventilation-i-bostader/

10

Biosorption of Heavy Metal by Algae Biomass in Surface Water

Handojo Djati Utomo[1*], Keng Xuan Donovan Tan[1], Zhi Yi Daniel Choong[1], Jia Jia Yu[1], Jie Jun Ong[1], Zheng Bang Lim[2]

[1]Division of Civil Engineering, School of Architecture and the Built Environment, Singapore Polytechnic, Singapore, Singapore
[2]Advanced Material Technology Centre, Technology Development Office, Singapore Polytechnic, Singapore, Singapore
Email: *han@sp.edu.sg

Abstract

Discharging wastewater containing heavy metals of Cu, Pb, Zn and Cd into water bodies can cause toxicity in plants and aquatic animals and some of them will be unable to survive except algae. Wastewater treatment method to remove heavy metal contaminants includes chemical precipitation, ion exchange, membrane, filtration, adsorption using activated carbon. However, these methods are either expensive or have other disadvantages such as high energy consumption and inefficiencies when existing heavy metals are at trace concentration. Biosorption using algae biomass can be an alternative method to eliminate heavy metals. The objective of the project is to investigate the capability of Marine Algae (MA) and Freshwater Algae (FA) biomass in adsorbing heavy metals of Cu, Pb, Zn and Cd from water medium using synthetic water and industrial water. MA and FA were obtained from the eastern coast of Pulau Ubin and local fish farm respectively. After being fully washed with deionised water, dried in a furnace for 105°C, they are grinded to pass 1 mm^2 of siever. MA and FA were characterised using FTIR to determine their functional groups. An industrial water was collected from industrial discharge from metal factories in northern side of Singapore. Effect of adsorption time, adsorbent concentration, and pH were studied. The result showed that FA and MA had a higher capability in adsorbing a total metal of about 40 ppm level from an industrial water, or 4 times than synthetic water concentration, at the same adsorbent dosage of 50 mg. In conclusion, the presence of various functional groups, hydroxyl, carboxylic and amine groups, in all MA and FA samples had enabled the algae biomass to adsorb heavy metals of Cu, Pb, Cd and Zn from synthetic and industrial water. Due to their biosorptive properties and fast adsorption capability, algae could be a potential method for cleaning up surface water or post-treatment of wastewater and minimise the cost of eutrophication.

Keywords

Algae Biomass, Biosorption, Eutrophication, Heavy Metal, Surface Water,

Wastewater

1. Introduction

Many countries have undergone rapid urbanisation and industrialisation over the last few decades. As a consequence, urban stormwater runoff increases directly with imperviousness and the degree of watershed development. Surface waters are forced to accommodate larger volumes of stormwater runoff that return on a more frequent basis. The change in watershed hydrology associated with urban development also causes channel widening and scour, and the introduction of larger amounts of suspended and dissolved sediments, such as heavy metals into river streams.

Heavy metals are toxic and will not only cause the pollution of water bodies but also potential public health problems and risks [1]. Hence, it is important to remove heavy metals such as Lead (Pb), Zinc (Zn), Copper (Cu) and Cadmium (Cd) which are common heavy metals that can be found in wastewater discharge. Lead, a naturally occurring metal found in the earth. Lead can infiltrate into the water supply through many means such as mining operations and lead water pipes. When lead is absorbed into the body, it causes risk of kidney damage and high blood pressure for adults. At high levels of exposure, lead attacks brain and nervous system causing coma, convulsion and even death. Copper sulfate can infiltrate the water supply through its use in mining industries. As copper sulfate is extremely soluble in water, it can easily seep into water bodies and supplies [2]. Persons suffering from copper poisoning suffer from gastrointestinal issues such as diarrhoea, nausea, abdominal pains and vomiting [3]. Consuming high concentration of copper sulfate can cause multiple organs to fail and death. Cadmium is a naturally occurring metal that can be found in metal ores and as natural deposits in the earth. Cadmium compounds are used in various industries such as pigments in plastics, an anticorrosive layer electroplated onto steel, electrical batteries and components. Cadmium can infiltrate the water supply through zinc pipes and solder containing cadmium that is used to supply water. High levels of cadmium could be found in shellfish and other filter feeding organisms. High amounts of cadmium can damage kidney and cause kidney stones to form and osteoporosis to occur [4]. Zinc can enter the water supply through pipe fittings or zinc pipes used to pipe water. Low pH in tap water can increase the amount of zinc leached into the water. High amounts of zinc can cause nausea, diarrhea, internal bleeding, vomiting and abdominal cramps [5].

Tannery, paper, pharmaceutical and food industries are some examples of industry that produces wastewater discharge with high concentration of heavy metal such as cadmium, lead and other toxic organics. Local authority regulates industries to comply with the standard limit for discharging wastewater into water bodies. However there are still some companies illegally discharging wastewater containing heavy metals and toxic compounds into the water bodies. Continuous incident in Singapore that caused million tonnes of aquatic organisms from rivers in Bishan-Ang Mo Kio, Sungei Buloh, Lim Chu Kang, extending to Changi and Pasir Ris Beach showed an extreme result of coastal water pollution. Many combinations of environmental factors can cause lack of oxygen and toxicity in water bodies. Plants and aquatic animals will be contaminated and some will be unable to survive except algae. As a consequence, the ability of wastewater treatment system to tolerate and remove toxicity is of considerable importance.

Algae biomass, in general, is an efficient adsorbent of heavy metals. Bio-treatment with microalgae is particularly attractive because of their photosynthetic capabilities, converting solar energy into useful biomasses and incorporating nutrients such as nitrogen and phosphorus causing eutrophication [6]. Also, algae based system for the removal of toxic minerals such as lead, cadmium, mercury, scandium, tin, arsenic and bromine are also being developed [7]-[11].

There are many ways to remove heavy metals from the wastewater, for example, chemical precipitation, ion exchange, membrane, filtration, adsorption using activated carbon and biological techniques [12]. However, these methods are either expensive, or have other disadvantages such as high energy consumption and inefficiencies when the existing heavy metals are in low concentration.

Biosorption can be an alternative method to eliminate heavy metals due to its low cost, high efficiency, minimization of secondary wastes and environmentally friendly. Green algae can be used as natural biosorbent to reduce chemical usage in waste treatment process which indirectly reduces the cost of the treatment plant.

Algae are defined as "thallophytes (plants lacking roots, stems, and leaves) that have chlorophyll as their primary photosynthetic pigment and lack a sterile covering of cells around the reproductive cells". The green algae are distributed in many regions of the world, from being airborne or subaerial to being terrestrial or aquatic, either in freshwater or marine habitats [13]. As an island surrounded by sea, Singapore's algae habitats are found in freshwater and marine water bodies. Therefore, in this project, marine algae (MA) and freshwater algae (MA) biomass are used as heavy metal scavenger in surface waters.

MA, commonly known as seaweed, is large enough to be seen and can grow up to 60 m in length. Types of seaweed include red, brown and green algae. MA do not have vascular system like plants, however they do have pigment chlorophyll for photosynthesis. In addition to the photosynthetic pigmentation, MA has many ultrastructural and biochemical features which include storage compounds, composition of cell walls, presence/absence of flagella, ultrastructure of mitosis, connections between adjacent cells, and fine structure of the chloroplasts [14].

FA is served as a paramount part of pond ecosystem, forming the base food chain. They provide food and energy for fish and other lake organisms and have the ability to absorb nutrients and heavy metals. The most common FA is known as cyanobacteria, contain bright, grass-green pigments. Generally most FA is highly palatable and a good food source for zooplankton [15].

The present study aims to investigate the capability of MA and FA in adsorbing heavy metals such as copper, lead, zinc and cadmium in water medium using spiked deionised water and real water sample from industrial discharge. Characterisation of MA and FA samples was conducted to identify the presence of various functional groups using FTIR analytical instrument. The concentration of all the above mentioned metals was measured using the latest technology of ICP-OES analytical instrument.

2. Methodology, Equipment and Materials

2.1. Preparation of MA and FA

An estimated mass of about 3 kg of MA was obtained directly from the eastern coast of Pulau Ubin as shown in **Figure 1(b)**. MA was thoroughly washed with tap water for an hour. This was carried out to remove sand and little cretaceous creatures. Foreign particle such as brown algae and dirt that was sticking to the MA was removed manually by hand. It was then soaked in deionized water for about an hour to ensure the cleanness of the samples before being placed in the oven at 104°C for 4- 6 hours to ensure that the samples are completely dried. All the dried algae was then grinded and sieved through a 1 mm^2 net to ensure that all the MA are smaller than 1 mm^2, thus forming a powdered MA. The grinded MA was then stored in labelled glass bottles. An estimated mass of about 8 kg of FA was obtained directly from local fish farm as shown in **Figure 1(a)**. FA was thoroughly washed with tap water to remove some particles such as dirt and leaves which stick and float respectively. Other smaller foreign particles such as duckweed and insects were removed using a tweezer. It was then processed following the above mentioned pre-treatment procedure for MA to form a powdered FA (**Figure 2(a)** and **Figure 2(b)**). The grinded FA was then stored separately in labelled glass bottles. The density of MA and FA were also calculated by weighing the mass of the material in 10 ml of 25 ml measuring cylinder.

(a) (b)

Figure 1. (a) Filtration system of fish farm and FA collection point (shown in red circle); (b) Shoreline of Pulau Ubin beach and MA collection point (shown on red circle).

(a) (b)

Figure 2. (a) Samples of grinded Freshwater Algae (FA); (b) Samples of grinded Marine Algae (MA).

2.2. Characterisation of MA and FA Using FTIR Analytical Instrument

MA and FA biomass characterisation were conducted using FTIR to determine the functional groups of MA and FA samples. No pre-treatment of the MA and FA samples was required due to the fact that the samples were already in powder form of biomass. As such it did not require the use of potassium bromide (KBr) pellets/method. The IR spectra of the samples are recorded by using IRPrestige-21 FTIR spectrophotometer. The spectra were recorded in the transmission band mode in the range of 750 - 4500 nm. Functional groups existing in MA and FA would be shown in the next result section *i.e.* before and after metal adsorption onto MA and FA surfaces.

2.3. Batch Adsorption Experiment Using Synthetic Heavy Metal Solution

0.05 g of FA or MA biomass powder was measured using Sartorius Quintix 224 1S Analytical Balance of accuracy up to four decimal places and was placed directly into the plastic spinning tube. A nominal 30 ml of deionized water was added to each of the plastic spinning tube. The plastic spinning tubes were placed onto the Rotary Mixing Solution YC-80 in the order of 60 min, 45 min, 35 min, 25 min, 15 min, 10 min and 5 min to determine the percentage of metal adsorption at various equilibrium times. Before the appropriate plastic spinning tubes were placed on the rotator, 0.075 ml of 1000 ppm Cu solution was added to each of the plastic spinning tubes to make 2.5 ppm metal concentration. The procedure was repeated for other metals of Pb, Cd, and Zn. All metal standard solutions were purchased from Sigma Aldrich, ICP Analytical Grade of 1000 ppm.

The tubes were rotated at 30 RPM at 45 degrees to ensure full contact and adequate mixing. Once the spinning was complete, the plastic spinning tubes were set to rest for 5 min to allow the MA and FA particles to settle in the bottom of tube. Once the settling time was complete, the solute was extracted and placed into two 15 ml plastic centrifuge tubes. The plastic centrifuge tubes were centrifuged for 3 minute at 6000 RPM to force the algae particle to the side of the plastic centrifuge tube separating the algae particles from the solution. Solution was extracted from the centrifuge tubes using droppers and placed into clean plastic spinning tubes. The remaining concentration of Pb, Cd, Cu, Zn ions in the solution were then measured using the ICP (Thermo Scientific iCAP 6000 Series ICP) at 220 nm, 228 nm, 324 nm, 213 nm respectively. The procedures were repeated twice for each metal ion.

The effect of adsorbent concentration was conducted by adding 0.5 mg, 5 mg, 25 mg, 50 mg, 100 mg, 200 mg and 300 mg added into 30 ml plastic tubes. 0.075 ml of metal standard solution with amount of deionized water was added into the plastic spinning tube, making up to 2.5 ppm of metal concentration. Rotation in the rotary mixer and centrifugation procedures were repeated, except a fixed rotation time of 60 mins. A longer rotation time was set up at 60 mins to fully ensure the metal adsorption process. In a subsequent batch experiment, the effect of pH was conducted by altering the pH of spiked solution with NaOH and HCl, ranging from pH 2 to 11. The abovementioned batch adsorption procedure was repeated, except a fixed rotation set up at 30 mins.

Various metals could exist in one single wastewater discharge. To mimic the real situation, a combined heavy metal was prepared in a subsequent experiment. The above-mentioned procedure was repeated using a combined four metal ions of Cu, Pb, Cd, Zn in a deionised water to understand the capacity and preferential metal adsorption onto MA and FA surfaces. Each metal was made 2.5 ppm, resulting a total metal concentration of 10 ppm from the four selected metals.

2.4. Batch Adsorption Experiment Using Real Industrial Water

A real industrial water was collected from industrial area of metal manufacturing factory. Fresh wastewater discharge after membrane wastewater treatment was collected using 1 L plastic bottles and stored and labelled in laboratory. Using the batch adsorption procedure in Subsection 2.3, all the metal concentration was measured using ICP-OES at once after two days of the settlement of larger particles. This is to reduce a filtration step process before measurement using sensitive equipment of ICP-OES, where the dissolved metals only could be measured.

3. Results and Discussion

3.1. MA and FA Biomass Characterisation Using FTIR

MA had a density 488.04 μg/mL, which was measured approximately 2.5 times denser than FA. **Figure 3** showed a scanned wave numbers from 4000 cm^{-1} to 1000 cm^{-1} from

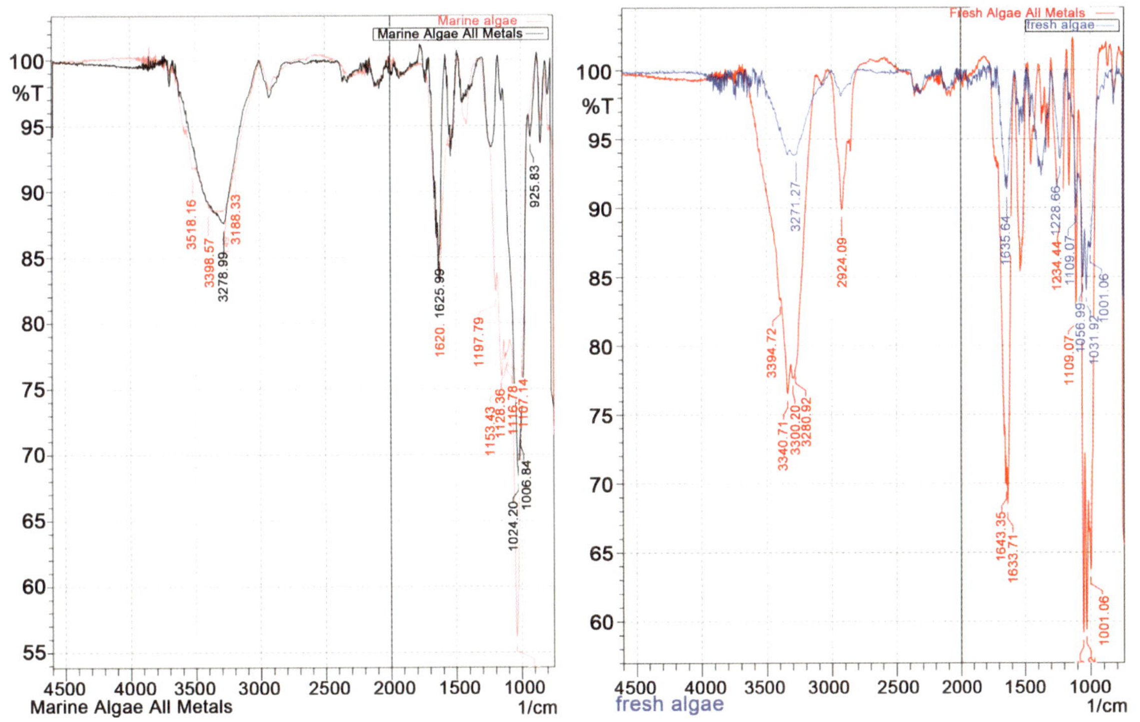

(a) (b)

Figure 3. (a), (b): FTIR spectra of MA (a-left) and FA (b-left) samples. Intensity changes are shown at different wave length before and after heavy metal adsorption onto MA and FA surfaces.

both FA and MA, before and after exposure to a 30 mL solution containing Cu, Pb, Zn and Cd at concentrations of 2.5 ppm. The FTIR analysis demonstrated which functional groups will be responsible for the adsorption of the metals present in the solution. **Figure 3(a)** shows FTIR spectra of MA sample with the bands associated with the hydroxyl O-H group, lignin and acetyl group. The bands at 3300 cm^{-1} and 2900 cm^{-1} are assigned to O-H stretching and C-H stretching respectively, and it exist in both MA and FA samples. MA had higher intensity at 3300 cm^{-1} (O-H stretching) than FA, however after all the four heavy metals adsorbed, FA showed the opposite result of higher O-H stretching. No significant intensity change after heavy metal adsorbed onto MA. The result showed that carboxylic functional group may be responsible for the heavy metal adsorption onto FA surface. A similar trend was also observed after the metals were bonded in the band 2900 cm^{-1} (C-H stretching). The band at 1600 cm^{-1} is assigned to C-C stretching in FA after the metals being adsorbed with no significant intensity change in MA. At the wavelength of 1000 cm^{-1} (C-O and C-N stretching), MA showed a large intensity change indicating that the heavy metals were adsorbed on the MA surface. The FA also showed similar significant results as MA. The result showed that the carboxylic and amine functional groups may be responsible for the heavy metal adsorption onto both MA and FA surfaces.

3.2. Effect of Adsorption Time

Figure 4 shows that FA was able to adsorb more Zn compared to other metals of Cu, Pb and Cd in separate solutions. The adsorption process showed a very fast adsorption process taking place in minutes and reaching a plateau after 20 minutes. Therefore in the subsequent analysis, effect of pH and concentration of adsorbent, a minimum fixed time of 30 mins would be used. FA adsorbed about 45 mg of Zn in each gram of FA. The reaction was instantaneous, removing almost 45 mg of Zn per gram of algae present

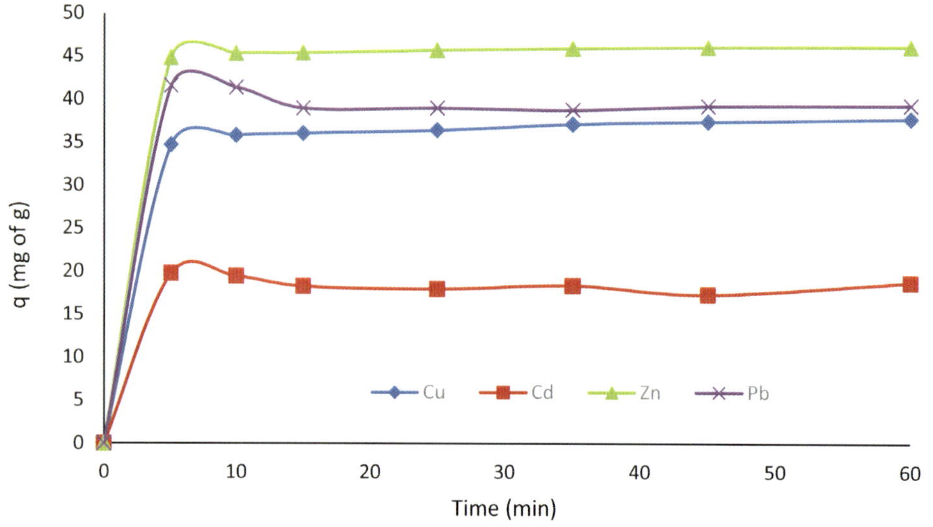

Figure 4. Adsorption kinetics of Cu, Pb, Cd, Zn onto by 0.05 g of FA. Initial concentration of metal is 2.5 ppm

within 5 min, it remains constant throughout the remaining adsorption time. FA was also able to adsorb Pb, Cu and Cd at lower concentration of 38 mg/g, 35 mg/g and 17 mg/g of FA respectively, measured after 30 mins. Alternatively, the percentage of adsorption onto 50 mg of FA could be calculated as 91.76%, 77.49%, 74.24% and 36.68% for Zn, Pb, Cu and Cd respectively.

Figure 5 shows MA was able to adsorb more Zn and Cu compared to other metals within 60 min. MA was able to adsorb about 42 mg/g of Zn within 60 min. It could also be observed that the percentage of heavy metal adsorbed remained constant throughout the entire 60 min. The reaction was also instantaneous adsorbing almost 43 mg/g of the Zn present within 5 min, the amount of heavy metal adsorbed started to decrease as the exposure time increased. The amount adsorbed decreased by about 1 mg/g. Since the adsorption did not have a significant change, it was considered that the reaction plateaued after 5 min. A similar trend was observed for Cu. As for Pb, MA was able to adsorb about 40 mg/g of the Pb within 60 min. However unlike the other metals, the amount adsorbed continued to increase at a steady rate as the exposure time increased, from 30 mg/g at 5 min to a stable rate of 40 mg/g at 60 min. Therefore the optimum time for Pb was considered after 60 mins. Cd is the least adsorbed metal by MA within 60 min, similar trend as observed for FA when adsorbing same Cd as the contaminant. MA was able to adsorb about 16 mg/g of Cd present within 60 min. The reaction was instantaneous, adsorbing 15 mg/g of the Cd present within 5 min. For simplicity, a fixed adsorption time of 60 minutes would be used in the subsequent experiments. Alternatively, the percentage of adsorption onto 50 mg of MA could be calculated as 84.52%, 84.38%, 78.84% and 33.24% for Zn, Cu, Pb and Cd respectively. The change of order of Cu and Pb metal ions adsorbed onto MA and FA showed that different functional groups, which exist on the surface of FA and MA, might have a preferential process of adsorption. Following the discussion in subsection 3.1, the result may be linked up that FA had a couple of functional groups, which may increase the heavy

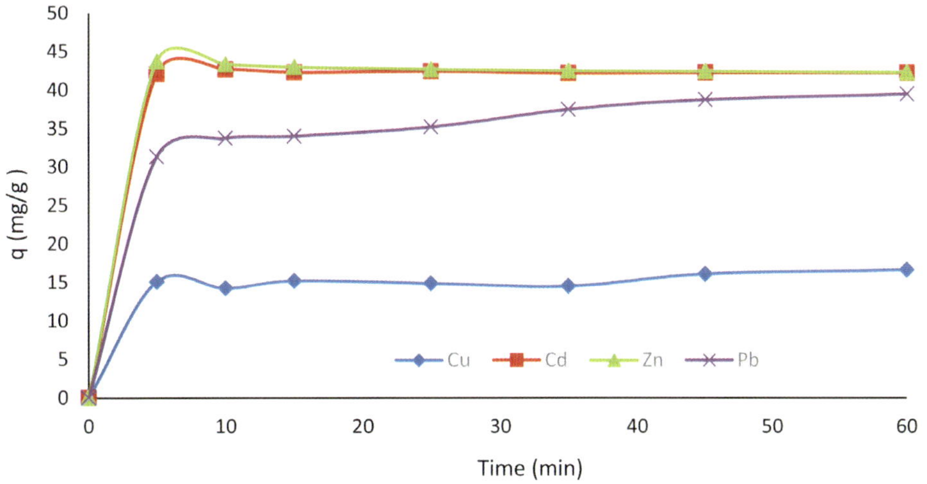

Figure 5. Adsorption kinetics of Cu, Pb, Cd, Zn onto by 0.05 g of MA. Initial concentration of metal is 2.5 ppm

metal adsorption rate. Meanwhile, MA had a singular large intensity change after the heavy metals being adsorbed at the wavelength of 1000 cm^{-1} due to carboxylic and amine functional groups.

Figure 6 shows the result of batch adsorption experiment using the combined four metals of Cu, Pb, Zn and Cd with the total metal concentration of 10 ppm. An instantaneous adsorption process was also observed in **Figure 7**. Pb is the most adsorbed metal compared to the other metals, followed by the Cu, Cd and Zn, which was the least. FA was able to adsorb about 23 mg/g, 10 mg/g, 9 mg/g and 7 mg/g of Pb, Cu, Cd and Zn respectively. FA's adsorption capability on each metal from the combined four metals solution showed a lower rate of about 50% - 70% than the previous FA's

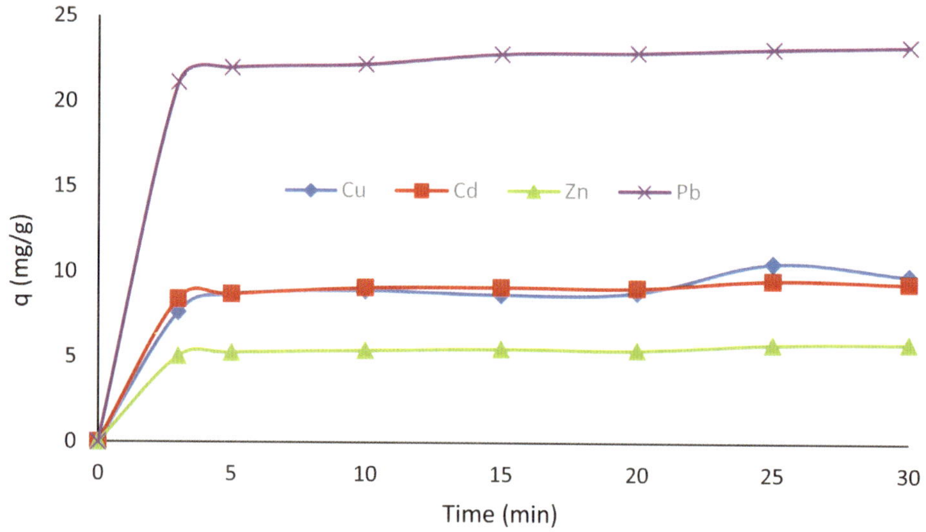

Figure 6. Adsorption kinetics of combined Cu, Pb, Cd, Zn onto by 0.05 g of FA. Total metal concentration is 10 ppm with equal concentration of each metal of 2.5 ppm.

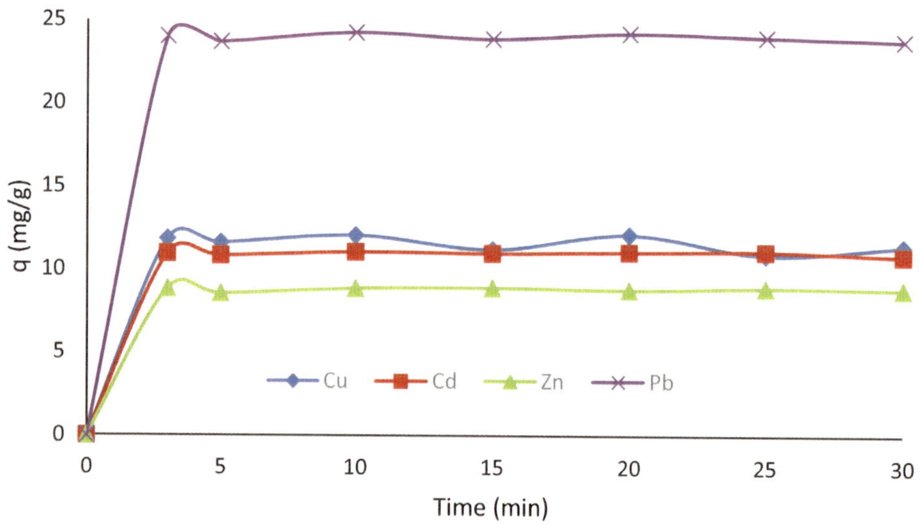

Figure 7. Adsorption kinetics of combined Cu, Pb, Cd, Zn onto by 0.05 g of MA. Total metal concentration is 10 ppm with equal concentration of each metal of 2.5 ppm.

adsorption capacity on each metal from individual metal solution. The result meant that the adsorption was technically limited to the availability of FA surface area with the active functional groups.

Figure 7 shows that similar fast adsorption process was also shown for MA with the order of adsorbed metals of Pb, Cu, Cd and Zn of 25 mg/g, 11 mg/g, 10 mg/g and 8 mg/g respectively. Since all the metal adsorption fluctuates at a difference of 1 mg/g, it could be considered constant. This means that MA achieved its maximum adsorption within 3 min of exposure, therefore the reaction was considered instantaneous. MA's adsorption capacity using combined four metals also showed a lower rate of about 50% - 70% than MA's adsorption capacity using individual metal.

The result showed a confirmation that adsorption process was limited to the availability of FA surface area with the active functional groups. **Figure 6** and **Figure 7** showed that both MA and FA were able to adsorb Pb more readily compared to the other metals and Zn is the least adsorbed metal.

3.3. Effect of MA and FA Adsorbent Concentration

An increasing amount of MA or FA may be able to fully adsorb heavy metals or reach an equilibrium state when reaching a plateau at a fixed concentration of metal. Percentage of each metal at fixed concentration of 2.5 ppm was adsorbed by FA and MA as presented in **Figure 8** and **Figure 9** respectively.

Figure 8 and **Figure 9** showed that the increasing mass of FA and MA were able to adsorb considerable percentage of metal in water. **Figure 8** showed that FA was able to adsorb Pb, Cd, Zn and Cu at nominal percentage of 94%, 75%, 86% and 53% respectively at 300 mg dosage. FA showed a lesser increment of percentage of adsorption on Cu rather than on Cd and Zn after 50 mg of FA. **Figure 9** showed that MA was able to adsorb Pb, Cu, Cd and Zn at nominal percentage of 92%, 64%, 61% and 53% respectively at 300 mg dosage. MA adsorbed all the four metals at a constant rate of adsorption

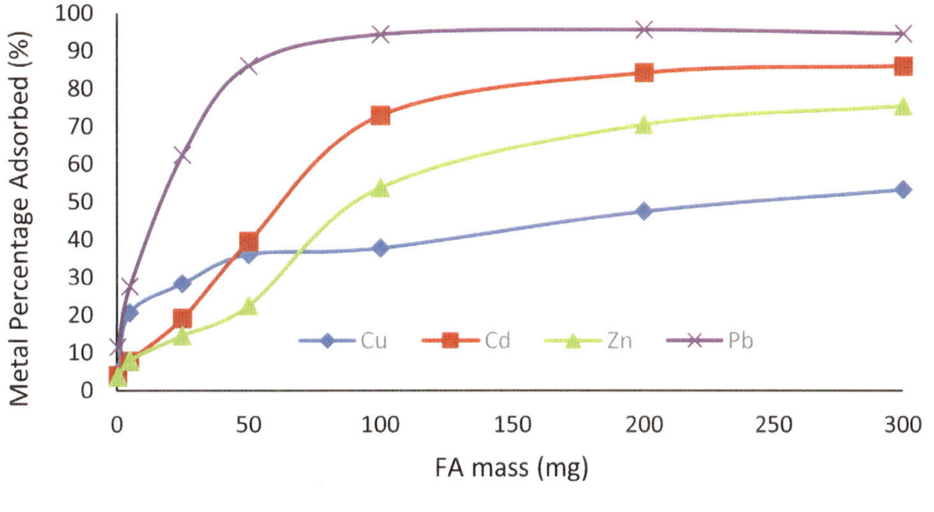

Figure 8. Percentage of metal of Cu, Pb, Cd and Zn onto FA. Initial concentration of each metal is 2.5 ppm.

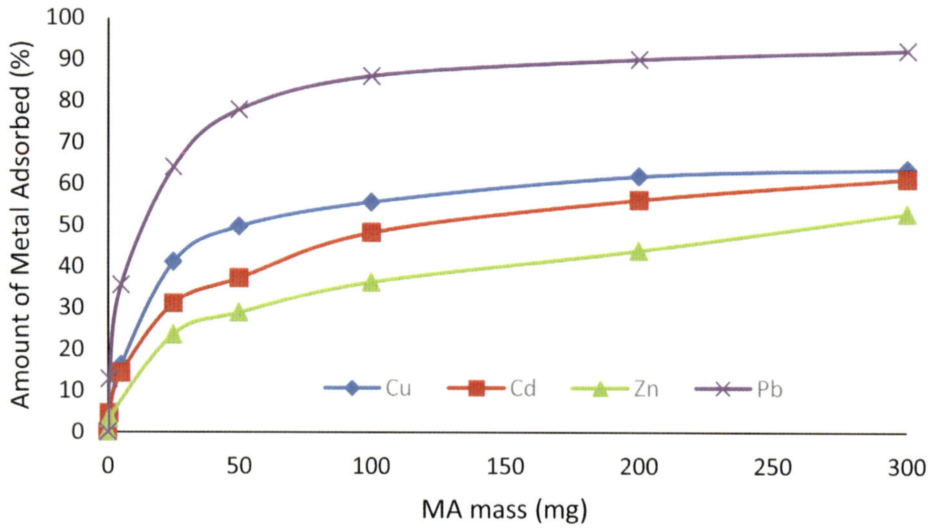

Figure 9. Percentage of metal of Cu, Pb, Cd and Zn onto MA. Initial concentration of each metal is 2.5 ppm.

until reaching plateau after 200 mg of MA dosage. Both FA and MA showed the dried algae have a strong percentage of adsorption on Pb at 94% and 92%.

3.4. Effect of pH on Heavy Metal Adsorption onto MA and FA

Figure 10 shows that the optimum pH for adsorption of heavy metals for FA is between pH of 5-pH of 10. It shows that the highest adsorption rates occur at pH 5 with the exception of Cd, which was most readily adsorbed at pH 11. It can be concluded that Pb was the highest adsorbed metal while Cu was the lowest adsorbed metal.

Figure 11 shows that the optimum pH for adsorption of heavy metals for MA also lies between pH 5 - pH 10. This is due to the plateau observed within the pH 5 - pH 10 range. Highest adsorption rates also occurred at pH 5. It is also concluded that Pb was the highest adsorbed metal while Zn was the lowest adsorbed metal.

3.5. Batch Adsorption Experiment Results Using Real Industrial Water

Figure 12 shows that the industrial water contained significant concentration of Cu and Zn dissolved ions, up to 10.425 ppm and 30.835 ppm respectively, and very minimum concentration of Cd and Pb metal ions. At a fixed equilibration time of 30 mins, MA showed a higher adsorption rate capacity for all metals than FA did. The result proved that FA and MA had a higher capability in adsorbing a total metal of about 40 ppm level from an industrial water, or 4 times than synthetic water concentration, at the same adsorbent dosage of 50 mg. Moreover, Zn was adsorbed at 6.87 mg/g and 7 mg/g of FA from an industrial water and synthetic water respectively. Similarly, Zn was also adsorb at 7.17 mg/g and 8 mg/g of MA from industrial water and synthetic water respectively. MA has higher adsorption of all metals when compared to FA. MA adsorbed 22%, 67%, 98%, 39% of Cd, Cu, Pb and Zn respectively, while FA adsorbed 18%, 29%, 94%, 37% of Cd, Cu, Pb and Zn respectively.

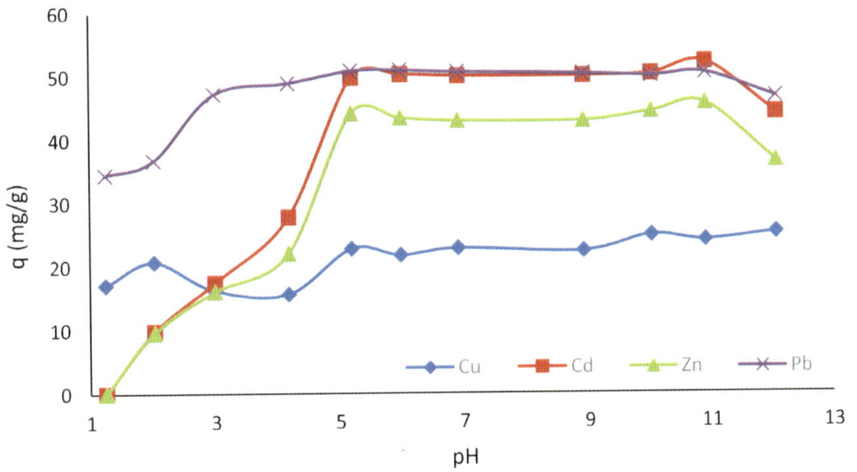

Figure 10. Adsorption rate of metal of Cu, Pb, Cd and Zn (mg/g) onto FA at a fixed time of 30 mins. Initial concentration of each metal is 2.5 ppm.

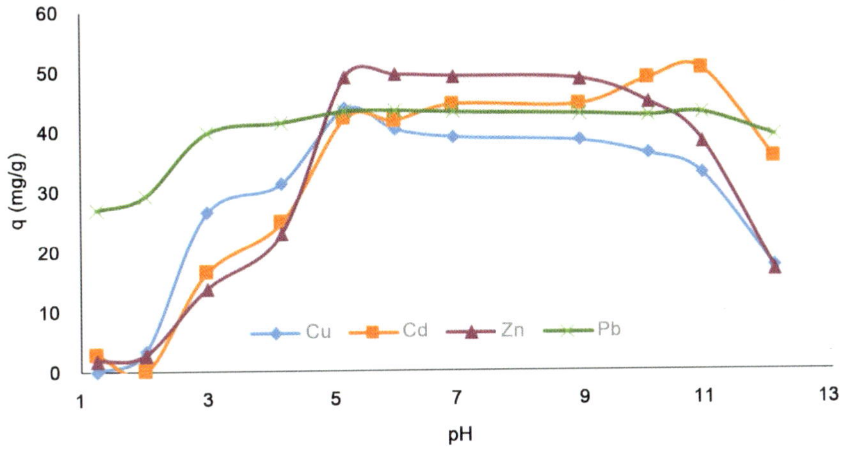

Figure 11. Adsorption rate of metal of Cu, Pb, Cd and Zn (mg/g) onto MA at a fixed time of 30 mins. Initial concentration of each metal is 2.5 ppm.

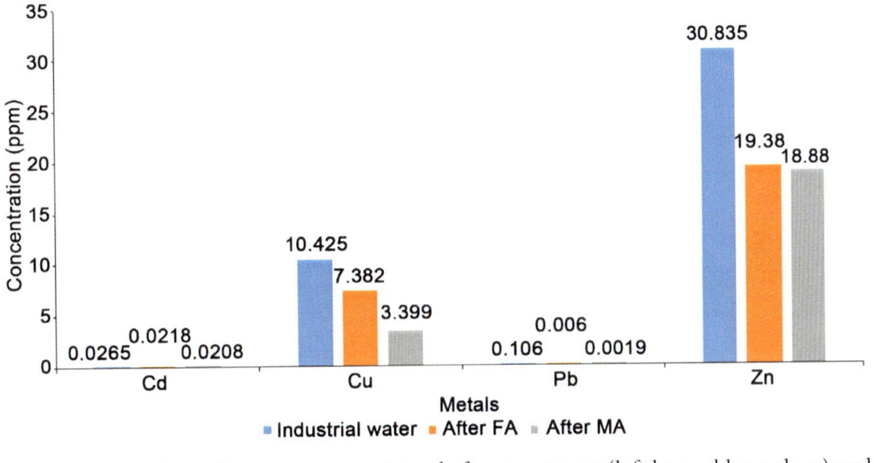

Figure 12. Comparison of metal concentration, before treatment (left bars—blue colour) and after treatment using FA (middle bars—orange colour) and MA (right bars—grey colour).

4. Conclusion

The presence of various functional groups, hydroxyl, carboxyl and amine groups, in all MA and FA samples had enabled the dried algae to adsorb heavy metals of Cu, Pb, Cd and Zn from industrial and synthetic water. Due to their biosorptive properties and fast adsorption capability, algae biomass could be a potential method for cleaning up surface water or post-treatment of wastewater. The MA and FA biomass have capability to adsorb heavy metals at intermediate pH, ranging from pH 5 to 11. Overall, use of algae biomass for heavy metal scavenger can offset the cost of eutrophication.

Acknowledgements

This project is supported by SP TIEFA (R& D) Fund with Vote Numbers 11-27801-36-R223.

References

[1] Freitas, O.M.M., Martins, R.J.E., Delerue-Matos, C.M. and Boaventura, R.A. (2008) Removal of Cd, Zn and Pb from Aqueous Solution by Brown Marine Macroalgae: Kinetic Modelling. *Journal of Hazardous Materials*, **153**, 493-501. http://dx.doi.org/10.1016/j.jhazmat.2007.08.081

[2] Gamakaranage, C.S.S.K., *et al.* (2011) Complications and Management of Acute Copper Sulphate Poisoning; A Case Discussion. *Journal of Occupational Medicine and Toxicology*, **6**, 34. http://dx.doi.org/10.1186/1745-6673-6-34

[3] Potera, C. (2004) Copper in Drinking Water: Using Symptoms of Exposure to Define Safety. *Environmental Health Perspectives*, **112**, A568. http://dx.doi.org/10.1289/ehp.112-a568b

[4] WHO (2001) Cadmium in Drinking Water, Background Document for Development of WHO Guidelines for Drinking-Water Quality. WHO/SDE/WSH/03.04/80/Rev/1.

[5] WHO (1996) Guidelines for Drinking-Water Quality. 2nd Edition, Vol. 2, Health Criteria and Other Supporting Information, World Health Organization, Geneva.

[6] De la Noue, J. and De Pauw, N. (1988) The Potential of Microalgal Biotechnology. A Review of Production and Uses of Microalgae. *Biotechnology Advances*, **6**, 725-770. http://dx.doi.org/10.1016/0734-9750(88)91921-0

[7] Soeder, C.J., Payer, H.D., Runkel, K.H., Beine, J. and Briele, E. (1978) Sorption and Concentration of Toxic Minerals by Mass Cultures of Chlorococcales. *Mitteilungen Internationale Vereinigung für The oretische und Angew and telimnologie*, **21**, 575-584.

[8] Kaplan, D., Christiaen, D. and Arad, S. (1988) Binding of Heavy Metals by Algal Polysaccharides. In: Stadler, T., Mollion, J., Verdus, M.C., Karamanos, Y., Morvan, H. and Christiaen, D., Eds., *Algal Biotechnology*, Elsevier Applied Science, London, 179-187.

[9] Gerhardt, M.B., Green, F.B., Newman, R.D., Lundquist, T.J., Tresan, R.B. and Oswald, W.J. (1991) Removal of Selenium Using a Novel Algal Bacterial Process. *Research Journal of the Water Pollution Control Federation*, **63**, 779-805.

[10] Hammouda, O., Gaber, A. and Abdel-Raouf, N. (1995) Microalgae and Wastewater Treatment. *Ecotoxicology and Environmental Safety*, **31**, 205-210. http://dx.doi.org/10.1006/eesa.1995.1064

[11] Hua, C.-X., Traina, S.J., Logan, T.J., Gustafson, T., Sayre, R.T. and Cai, X.H. (1995) Applications of Eukaryotic Algae for the Removal of Heavy Metals from Water. *Molecular Ma-*

rine Biology and Biotechnology, **4**, 338-344.

[12] Baraka, M.A. (2010) New Trends in Removing Heavy Metals from Industrial Wastewater. *Arabian Journal of Chemistry*, **4**, 361-377. http://dx.doi.org/10.1016/j.arabjc.2010.07.019

[13] Lee, A.C. and Liao, L.M. (2009) New Records of Marine Algae on Artificial Structures and Intertidal Flats in Coastal Waters of Singapore. *Raffles Bulletin of Zoology*, **22**, 5-40.

[14] Seaweed (2016) What Are Seaweeds? http://www.seaweed.ie/Algae/seaweeds.php

[15] United State Geological Survey (1978) Primer on Lakes in Washington (Water Supply Bulletin). United States Department of the Geological Interior Survey, Washington, 49.

11

Investigation on ^{226}Ra, ^{238}Th, ^{40}K and ^{137}Cs Concentrations in Common Polishing Materials Consumed by Inhabitants in Saudi Arabia

Wedad Rayif Alharbi

Department of Physics, Faculty Science, King Abdulaziz University, Jeddah, Saudi Arabia
Email: walharbi@kau.edu.sa

Abstract

Knowledge of radioactivity present in polishing materials enables one to assess any possible radiological risks to human health. In this work, the radioactivity due to the presence of ^{226}Ra, ^{232}Th, ^{40}K, and ^{37}Cs has been measured in polishing materials consumed in Saudi Arabia using gamma spectrometry with HPGe. The activity concentrations of ^{226}Ra, ^{232}Th, ^{40}K and ^{137}Cs ranged from 13.61 ± 0.005 to 0.60 ± 0.002, 18.43 ± 0.003 to 0.78 ± 0.001, 342.59 ± 0.009 to 2.47 ± 0.001 and 1.47 ± 0.001 to 0.55 ± 0.001 Bq/l, respectively. For ^{226}Ra and ^{40}K, the highest values are measured in stainless steel polisher. The highest values also measured in metal polisher (copper-chrome) and disinfectant cleaner and polisher sample for ^{232}Th and ^{137}Cs. Radium equivalent activity due to the natural radioactivity of the investigated samples ranged from 51.37 to 1.24 Bq/l. This value is less than the recommended values in the established standards. The evaluated data were compared with the literature data. Our results indicate that no significant radiological hazards arise from using investigated samples hence quite safe to be used as polishing materials.

Keywords

Gamma Spectrometry, Radiation Hazard, Polishing Materials, Natural Radioactivity

1. Introduction

Naturally occurring radioisotopes (^{238}Th, ^{232}U series and ^{40}K) are the main source of both external and internal radiation exposure in humans, whence the radiation finds its way into building materials, air, water, food, and eventually the human body. Human beings are exposed to external radiation from cosmic rays and terrestrial radiation. These isotopes enter the human body by intake from food and inhalation and result in exposure to internal radiation [1] [2]. According to the report by UNSCEAR, the total exposure per person resulting from ingestion of terrestrial radioisotopes was 0.3 mSv, of which 0.17 mSv comes from ^{40}K and 0.12 mSv comes from thorium and uranium series. Exposure from inhalation of terrestrial radioisotopes contributes another 0.01 mSv. Thus, the internal dose of terrestrial radioisotopes can be estimated from the concentrations of natural radioisotopes in foods [3]. Internal radiation exposure, mainly affecting various levels of the respiratory tract, is due to the deposition of short lived progeny products of radon, which are exhaled from building material into indoor air [4] [5]. Radon (^{222}Rn), a gaseous product of decay of ^{226}Ra, is extremely important in an indoor exposure. Radon easily diffuses from the ground and building materials to indoor air and is the source of exposing bronchus and lungs [6]-[8]. The isotope ^{137}Cs is produced anthropogenically by several types of nuclear activity, including past testing of nuclear weapons, accidents in nuclear facilities, reprocessing of spent nuclear fuel and nuclear power reactors. All these materials contain some amount of natural radionuclides that cause exposure of people to ionizing radiation. The measurement of activity concentrations of radionuclides in polishing materials is important in the assessment of population exposures, as most individuals spend 80% of their time indoors [9]. Long exposures to low levels of ionizing radiation and poor ventilation of residential buildings can seriously increase health risks to polishing materials [10]. Polishing materials have a variety of chemical natures and some of them are toxic to humans and may cause sickness including cancers and death [11]. Even nontoxic polishing materials can pollute the land or water. Therefore, care needs to be exercised when using and disposing such materials [12]. The present study was undertaken with the purpose of determining natural radioactivity in some polishing materials. The data obtained are essential for the development of standards and guidelines concerning the use and management of polishing materials.

2. Sampling and Measurements

Twenty-two samples representing commercial species of the most commonly consumed polishing materials were collected from Saudi Arabia local market. They were being used for cleaning and polishing purposes of marble, granite, ceramic, tiles, stucco, brick, porcelain, chrome and metal, houses and vehicles. Weighed samples were stored in standard polyethylene Marinelli beakers of 650 cm^3 volume each. The beakers were tightly sealed for 4 - 5 weeks to reach secular equilibrium where the rate of decay of the progeny becomes equal to that of the parent (radium and thorium) [13]. **Table 1** shows samples description.

3. Gamma Spectrometric Measurements

The activity concentrations of ^{226}Ra, ^{232}Th, ^{40}K and ^{137}Cs were measured with gamma-spectrometric system based on HPGe. The samples were counted using a gamma spectrometry (Canberra coaxial-type Model GC2520) with relative efficiency of 25% and resolution FWHM of 0.2 keV for 1332.5 keV gamma-ray peak of ^{60}Co and a peak to

Table 1. Sample code and their description.

Sample code	Sample description	Country
P1	Marble and ceramic cleaner	Saudi Arabia
P2	Super thick cleaner with Bleach	Saudi Arabia
P3	Ceramic cleaner, disinfectant and polisher	Saudi Arabia
P4	Eliminator bad odors cleaner and polisher	Italy
P5	Stainless steel cleaner and polisher	U.S.A
P6	Metal polisher (copper-chrome)	Nederland
P7	Antiseptic disinfectant	U.A.E
P8	Disinfectant, cleaner and polisher	Saudi Arabia
P9	Super polisher for all paint types including metallic	England
P10	High glosser, protective paintwork sealant for the perfectionist	England
P11	Rapidly remover for intensive tar, adhesive, grease, wax, oil and silicone	Britain
P12	Silver polish	Italy
P13	Tiles and surface cleaner(cement, gypsum, rust and dirt remover	Saudi Arabia
P14	Instant disinfectant, cleaner and polisher pine scent	U.A.E
P15	Instant disinfectant, cleaner and polisher pine scent	Saudi Arabia
P16	Cleaner and polisher	U.S.A
P17	Advanced cleaner and polisher	U.S.A
P18	Multi-surface cleaner and disinfectant—powerful scent	U.S.A
P19	Windows, mirrors, appliances, counters, Tiles and other smooth surface cleaner and polisher	Saudi Arabia
P20	Lime solvent from pots, tube and sinks rust remover, stains from stucco, brick, porcelain, chrome and metal-blasts calcium from glassware and tile	U.S.A
P21	Glass and household cleaner and polisher	Saudi Arabia
P22	Concentration cleaner and polisher of marble, granite, ceramic and tiles	Spain

Compton ratio of 50:1. The gamma-spectrometry system consists of HPGe connected to a desk top computer provided with a Canberra multichannel analyzer (MCA) in conjunction with a configuration software for spectrum acquisition and evaluation [14] [15].

The specific activities were averaged from gamma-ray photo peaks at several energies. The gamma-ray lines at 295.2 and 351.9 keV from ^{214}Pb and at 609.3 and 1764.5 keV from ^{214}Bi were used to determine the specific activity of ^{226}Ra. The gamma ray lines of 338.4 keV; 911.2 keV and 968.97 keV from ^{228}Ac, the 238.58 keV and 727.3 keV from ^{212}Bi and 583.2 keV from ^{208}Tl were used to determine the specific activity of ^{232}Th. The specific activities of ^{40}K and ^{137}Cs were measured directly by its own gamma-ray lines at 1460.8 and 661.66 keV, respectively [13]. The minimum detectable amount (BDLs) for ^{226}Ra, ^{232}Th, ^{40}K and ^{137}Cs were 0.312 ± 0.16, 0.340 ± 0.16, 1.660 ± 1.10 and

0.45 ± 0.06 Bq/l respectively. Radioactivity concentrations of each sample were measured for about 23 h. The data acquisition, display and spectrum analysis were carried out using a dedicated software program [16], which enabled also the concentrations of ^{226}Ra, ^{232}Th and ^{40}K to be calculated. The degree of secular equilibrium reached between ^{226}Ra and ^{232}Th and their decay products was taken into account during the concentration calculations.

4. Results and Discussion

The specific activity of ^{226}Ra, ^{232}Th, ^{40}K and ^{137}Cs, in the samples are shown in **Table 2.** The highest average concentrations of ^{226}Ra, ^{232}Th, ^{40}K and ^{137}Cs (13.60 ± 0.0015, 18.43 ± 0.0031, 342.59 ± 0.0091and 1.47 ± 0.0005 Bq/l respectively) were found in samples P5,

Table 2. Average of ^{226}Ra,^{232}Th, ^{40}K and Ra_{eq} of polishing materials in the current study.

Sample code	Activity concentration ($Bq{\cdot}l^{-1}$)				Ra_{eq} ($Bq{\cdot}l^{-1}$)
	^{226}Ra	^{238}Th	^{40}K	^{137}Cs	
P1	1.55 ± 0.001	1.56 ± 0.004	61.46 ± 0.001	1.06 ± 0.001	8.52
P2	1.23 ± 0.001	5.19 ± 0.006	39.26 ± 0.001	ND	11.67
P3	0.60 ± 0.002	2.10 ± 0.005	ND	0.96 ± 0.001	3.71
P4	ND	1.70 ± 0.003	ND	0.72 ± 0.001	2.46
P5	13.61 ± 0.005	17.48 ± 0.011	342.59 ± 0.009	0.55 ± 0.001	51.37
P6	9.50 ± 0.003	18.43 ± 0.003	36.34 ± 0.001	0.98 ± 0.001	38.66
P7	2.82 ± 0.003	2.60 ± 0.001	40.64 ± 0.001	0.75 ± 0.001	9.65
P8	1.02 ± 0.002	5.24 ± 0.007	10.04 ± 0.001	1.47 ± 0.001	9.28
P9	4.22 ± 0.001	11.23 ± 0.005	11.74 ± 0.001	1.19 ± 0.001	21.18
P10	2.96 ± 0.002	2.24 ± 0.001	6.13 ± 0.001	0.63 ± 0.001	6.64
P11	2.27 ± 0.003	1.90 ± 0.003	5.63 ± 0.001	ND	5.43
P12	11.30 ± 0.001	3.92 ± 0.004	31.93 ± 0.001	ND	19.36
P13	0.67 ± 0.009	3.15 ± 0.005	ND	ND	5.17
P14	2.12 ± 0.006	ND	4.05 ± 0.001	ND	2.93
P15	2.27 ± 0.006	3.28 ± 0.005	2.47 ± 0.001	ND	7.16
P16	4.09 ± 0.005	2.84 ± 0.003	ND	ND	8.15
P17	ND	0.78 ± 0.001	ND	ND	1.24
P18	1.77 ± 0.005	1.60 ± 0.004	ND	ND	5.58
P19	0.79 ± 0.002	1.16 ± 0.001	ND	ND	2.46
p20	1.35 ± 0.004	1.33 ± 0.004	ND	ND	3.25
P21	1.94 ± 0.002	2.25 ± 0.003	48.88 ± 0.001	ND	8.92
P22	2.60 ± 0.002	3.70 ± 0.004	57.14 ± 0.007	ND	12.30

ND: Not detected.

P6, P5 and P8 respectively, whereas the lowest average concentrations ^{226}Ra, ^{232}Th, ^{40}K and ^{137}Cs (0.60 ± 0.0015, 0.78 ± 0.0011, 2.47 ± 0.0005, 0.55 ± 0.0008 Bq/l respectively) were found in samples P3, P17, P15 and P5. The ^{40}K activity concentration dominated over that of the ^{226}Ra and ^{232}Th, as normally happens in samples, while the highest activity level of ^{40}K (342 Bq/l) which was recorded in sample P5 was lower than that of the global average of 400 $Bq{\cdot}kg^{-1}$ [17].

The variation among the activity levels in different samples may be attributed to the pH, and chemical composition of the raw materials from which they derived [18]. Some studies present the concentration values of ^{226}Ra only and do not consider the ^{238}U presence taking into account the fact that 98.5% of the radiological effects of the uranium series are produced by radium and its daughters [19].

The activity concentrations of Ra, presented here, fall within the range reported by other authors. Both Th and K in this work revealed higher concentration levels compared to those level for painting oxides and paints. However, their levels were lower than in the case of titanium enamel frits (**Table 3**). According to the recommended reference level of 30, 25 and 370 Bq/kg for ^{226}Ra, ^{232}Th and ^{40}K, respectively, for the world average concentrations published by UNSCEAR [1], it was noted that the obtained results in all samples are lower than the recommended reference level.

5. Radiological Parameters

Radium Equivalent Activities (Ra_{eq})

Uniformity of the distribution of ^{226}Ra, ^{232}Th and ^{40}K in materials with respect to radiation exposure has been described by a common index. This index is called the radium equivalent (Ra_{eq}) activity and is defined as a weighted sum of the activity concentrations of the above three radionuclides. Ra_{eq} is given in $Bq{\cdot}l^{-1}$ to compare the specific activity of materials containing different amounts of ^{226}Ra, ^{232}Th and ^{40}K. It was assumed that 370 $Bq{\cdot}kg^{-1}$ of ^{226}Ra, 259 $Bq{\cdot}kg^{-1}$ of ^{232}Th or 4810 $Bq{\cdot}kg^{-1}$ of ^{40}K produce the same γ-ray dose rates [20]; Ra_{eq} is calculated through the following relationship:

$$Ra_{eq} = C_{Ra} + 1.43C_{Th} + 0.077C_K \quad (1)$$

where C_{Ra}, C_{Th}, and C_K are the specific activities of ^{226}Ra, ^{232}Th, and ^{40}K, respectively, expressed in $Bq{\cdot}l^{-1}$. The maximum value of Ra_{eq} in building materials must be less than 370 $Bq{\cdot}kg^{-1}$ for safe use of materials in the construction of buildings [1], to keep the external dose below 1.5 $mSv{\cdot}y^{-1}$ [21]. The results obtained by us showed that the lowest

Table 3. Comparison of concentrations range of ^{226}Ra, ^{232}Th, and ^{40}K ($Bq{\cdot}kg^{-1}$) recorded in various polishing materials of several studies.

Materials	Activity concentration			References
	^{226}Ra	^{232}Th	^{40}K	
Titanium enamel frits	1.05 - 1018.57	1.05 - 1164.09	942.29 - 16,679.19	[22]
Painting oxides	2.35 - 72.96	1.76 - 12.88	2.26 - 200	[23]
Paits	10.5 - 43.0	0.5 - 0.5	0.5 - 7.7	[24]
Polishing	0.60 - 13.61	0.78 - 18.43	2.47 - 342.59	Present work

Ra_{eq} was equal to 1.24 $Bq \cdot l^{-1}$ in sample P17, while the highest value reached 51.37 $Bq \cdot l^{-1}$ in in sample P5 (**Table 2**). All the studied samples showed lower values than the recommended worldwide mean value 370 Bq kg^{-1}. Thus, every material used to clean and polish of marble, granite, ceramic and tile can be considered to does not reveal a significant radiological hazard.

6. Conclusion

The activity levels of the natural terrestrial radionuclides as ^{226}Ra, ^{232}Th, ^{40}K and artificial ^{137}Cs were determined using a gamma-ray spectrometry in 22 samples of polishing materials collected from local stores. The obtained values were, in general, comparable to the corresponding ones obtained from other studies, and they all fell within the average worldwide ranges. Basing on the measurement results, the values of the radium equivalent activities Ra_{eq} were calculated to assess the radiological hazard for inhabitants of Jeddah. The obtained values displayed to be below the permissible limiting values.

References

[1] UNSCEAR (2000) Sources and Effects of Ionizing Radiation. Report to General Assembly, Annex B Exposure from Natural Radiation Sources, United Nations, New York.

[2] Maxwell, O., Wagiran, H., Lee, S.K., Embong, Z. and Ugwuoke, P.E. (2015) Radioactivity Level and Toxic Elemental Concentration in Ground Water at Dei-Dei and Kubwa Areas of Abuja, North-Central Nigeria. *Radiation Physics and Chemistry*, **107**, 23-30. http://dx.doi.org/10.1016/j.radphyschem.2014.09.003

[3] Faanu, A., Adukpo, O.K., Tettey-Larbi, L., *et al.* (2016) Natural Radioactivity Levels in Soils, Rocks and Water at a Mining Concession of Perseus Gold Mine and Surrounding Towns in Central Region of Ghana. *Springerplus*, **5**, 98. http://dx.doi.org/10.1186/s40064-016-1716-5

[4] El-Taher, A. and Madkour, H.A. (2014) Environmental Studies and Radio-Ecological Impacts of Anthropogenic Areas: Shallow Marine Sediments Red Sea, Egypt. *Journal of Isotopes in Environment and Health Studies*, **50**, 120-133. http://dx.doi.org/10.1080/10256016.2013.826211

[5] Stoulos, S., Manolopoulou, M. and Papastefanou, C. (2003) Assessment of Natural Radiation Exposure and Radon Exhalation from Building Materials in Greece. *Journal of Environmental Radioactivity*, **69**, 225-240. http://dx.doi.org/10.1016/S0265-931X(03)00081-X

[6] UNSCEAR (1988) United Nations Scientific Committee on the Effects of Atomic Radiation. Sources, Effects and Risks of Ionizing Radiation. Report to the General Assembly, 32nd Session, United Nations, New York, 183-184.

[7] Stidley, C.A. and Samet, J.M. (1993) A Review of Ecological Studies of Lung-Cancer and Indoor Radon. *Health Physics*, **65**, 234-251. http://dx.doi.org/10.1097/00004032-199309000-00001

[8] Alharbi, W.R. and Abbady, A.G.E. (2013) Measurement of Radon Concentrations in Soil and the Extent of Their Impact on the Environment from Al-Qassim, Saudi Arabia. *Natural Science*, **5**, 93-98. http://dx.doi.org/10.4236/ns.2013.51015

[9] UNSCEAR (1993) Sources and Effects of Ionizing Radiation. United Nations Scientific Committee on the Effect of Atomic Radiation.

[10] Guidotti, L., Carini, F., Rossi, R., Gatti, M., Cenci, R.M. and Beone, G.M. (2015) Gamma-

Spectrometric Measurement of Radioactivity in Agricultural Soils of the Lombardia Region, Northern Italy. *Journal of Environmental Radioactivity*, **142**, 36-44. http://dx.doi.org/10.1016/j.jenvrad.2015.01.010

[11] El Aal, S.A., Korman, A., Stonert, A., Munnik, F. and Turos, A. (2009) Ion Beam Analysis of Ancient Egyptian Wall Paintings. *Vacuum*, **83**, 54-58. http://dx.doi.org/10.1016/j.vacuum.2009.01.012

[12] Adebamowo, E.O., Agbede, O.A., Sridhar, M.K.C. and Adebamowo, C.A. (2006) An Evaluation of Lead Levels in Paints for Residential Use Sold in the Nigerian Market. *Indoor and Built Environment*, **15**, 551-554. http://dx.doi.org/10.1177/1420326X06072992

[13] Yussuf, N.M., Hossain, I. and Wagiran, H. (2012) Natural Radioactivity in Drinking and Mineral Water in Johor Bahru (Malaysia). *Scientific Research and Essays*, **7**, 1070-1075.

[14] Sharma, N., Singh, J., Esakki, S.C. and Tripathi, R.M. (2016) A Study of the Natural Radioactivity and Radon Exhalation Rate in Some Cements Used in India and Its Radiological Significance. *Journal of Radiation Research and Applied Sciences*, **9**, 47-56.

[15] Challan, M.B. and El-Taher, A. (2014) Analytical Approach for Radioactivity Correlation of Disc Sources with HPGe Detector Efficiency. *Journal of Applied Radiation and Isotopes*, **85**, 23-27. http://dx.doi.org/10.1016/j.apradiso.2013.11.087

[16] GENIE-2000 (1997) Basic Spectroscopy (Standalone) V1.2A Copyright (c), Canberra Industries.

[17] Mehra, R. (2009) Use of Gamma Ray Spectroscopy Measurements for Assessment of the Average Effective Dose from the Analysis of ^{226}Ra, ^{232}Th, and ^{40}K in Soil Samples. *Indoor Built Environ*, **18**, 270-275. http://dx.doi.org/10.1177/1420326X09104140

[18] Shanthi, G., Maniyan, C.G., Allan Gnana Raj, G. and Thampi Thanka Kumaran, J. (2009) Radioactivity in Food Crops from High Background Radiation Area in Southwest India. *Current Science*, **97**, 1331-1335.

[19] Shousha, H.A. (2006) Radioactive Analysis and Radiological Hazards in Different Types of Egyptian Cement. *Radiation Effects & Defects in Solids*, **161**, 615-627. http://dx.doi.org/10.1080/10420150600858371

[20] Beretka, J. and Mathew, P.J. (1985) Natural Radioactivity of Australian Building Materials, Industrial Wastes and By-Products. *Health Physics*, **48**, 87-95. http://dx.doi.org/10.1097/00004032-198501000-00007

[21] NEA-OECD (1979) Nuclear Energy Agency, Exposure to Radiation from Natural Radioactivity in Building Materials. Nuclear Energy Agency (NEA), Report by NEA Group of Experts, Organization for Economic Co-Operation and Development, OECD, Paris, France.

[22] Amekudzie, A., Adukpo O., Annkah, J., Faanu, A. and Emi-Reynolds, G. (2011) Determination of ^{40}K, ^{226}Ra and ^{232}Th Concentration and Radiation Dose Assessment in Titanium Enamel Frits Use in Polishing Household Utensils in Ghana. *Journal of Radioanalytical and Nuclear Chemistry*, **289**, 795-800. http://dx.doi.org/10.1007/s10967-011-1177-9

[23] Abel-Ghany, H.A. (2011) Radiation Hazard Assessment in Egyptian Painting Oxides. A Comparative Study. *Environmental Geochemistry and Health*, **33**, 225-234. http://dx.doi.org/10.1007/s10653-010-9336-4

[24] Ali, K.K. (2012) Radioactivity in Building Materials in Iraq. *Radiation Protection Dosimetry*, **148**, 372-379. http://dx.doi.org/10.1093/rpd/ncr033

12

Visual Impact Assessment: Where have we Come from and Where are we Going?*

Richard C. Smardon

Department of Environmental Studies, SUNY College of Environmental Science and Forestry, Syracuse, USA
Email: rsmardon@esf.edu

Abstract

This paper is a historical overview of the development of visual impact assessment (VIA) methodology as implemented by US Federal agencies from the 1960s until the current time. VIA methods are to be used as part of environmental impact assessment for projects which could have significant impact on landscape aesthetics. Also included are methodological challenges with current large-scale renewable energy development in the US and Europe.

Keywords

Visual Impact Assessment, Methodology History, Federal Agencies, Challenges

1. Introduction

In the 1960's and 1970's there were many funded and/or permitted federal projects that were at a standstill because of lack of public involvement and/or incorporation of environmental and social values [1] & [2]. With the passage of the National Environmental Policy Act (NEPA) in 1970 [3] there now had to be incorporation of environmental values into the decision making process. Aesthetics is one of these values so the 1970's and 1980's saw an evolution and development of visual landscape management systems by a number of Federal agencies [1] [2] & [4]. This paper will describe this evolution of visual/landscape/scenic management systems from the 1970's to current time with particular emphasis on development of "defensible" visual impact assessment (VIA) methods.

The US Forest Service [5], USDI Bureau of Land Management [6] and the Soil Conservation Service [7], which became the Natural Resources Conservation Service

*An earlier version of this paper was presented at the 39th Annual National Association of Environmental Professionals Conference, Changing Tides and Shifting Sands April 7-10, 2014, St. Petersburg, Florida.

(NRCS), were the early leaders among US Federal Agencies.

The US Forest service faced blockage of a number of timber management plans in 1970's due to lack of public input and consideration of environmental values such as wildlife habitat and aesthetics within the review process, particularly in the Pacific Northwest [1] & [2]. Bacon, Orr and other landscape architects developed the early Visual Management System with ideas from several US Forest Service regions [2]. In 1972 the US Forest Service in Washington DC backed the development of this same system and decided to implement this system [8] & [9] within two forest service regions in the Pacific Northwest. Further input and review was received from R. Burton Litton, Jr. who produced one of the first Forest Service method papers on visual control points in 1973 [10]. The Forest Service Visual Management System (VMS) was put in US Forest Service regulations in 1976 [11] and was gradually implemented throughout many Federal Forests as part of timber management and land use management planning from 1975 to 1976 and thereafter [2] & [5]. In 1995 the US Forest Service updated this system to a Scenic Management System developed by Lee Anderson [12]. Both systems include landscape classification, visibility of activities to sensitive users, and ability of the landscape to absorb visual impact. A series of handbooks were developed after 1976 to address the potential impacts of various activities such as timber management, utility lines and recreational development.

The USDI Bureau of Land Management (BLM) has purview over grazing permits on Federal lands, some forestry activities, oil and gas development, and mining activity (until recently). In the mid 1970's Robert Leopold, head landscape architect for BLM, with the input from some that worked on the Forest Service's VMS developed their own VMS. This system included landscape classification of scenic attributes, visibility distance zones and a visual impact assessment (VIA) methodology [13]. This VIA methodology was called a visual contrast rating system, which evaluated the degree of contrast of the introduced project to existing landforms, vegetation and land use via shifts in line, form color and texture. Training sessions were held throughout the BLM districts including the first computer graphics session in 1978. The official VMS was published in 1980 as a two-volume set-one on the VMS itself [14] and one on simulation methods [15]. The author worked on a multidisciplinary team with Nick Feimer and Stephen Sheppard [16] at the University of California, Berkeley from 1977 to 1979 to test the reliability, validity and generalizability of the contrast rating methodology. One of the major findings from this study is that multiple raters (professional or trained) are needed to ensure reliability in use of such systems as well as detailed guidance for use of such systems to increase both reliability and validity [17] & [18]. There have been few studies since [19]-[21] which have looked at validity and reliability issues of doing VIA work.

The Soil Conservation Service (SCS), which is now the Natural Resource Conservation Service (NRCS), developed their own landscape assessment method in 1972 with the hiring of Sally Schauman-their first landscape architect [2] & [7]. SCS/NRCS, unlike the Forest Service or BLM, does not manage federal lands, but offers technical support programs and funding to private landowners to encourage better environmental practices. They also need to do environmental assessments for such federally funded projects as well. Under Sally Schauman's direction in 1978-SCS/NRCS published the in-house document TR-65 [22], which provides guidance for assessment of landscape attributes, impacted by proposed projects. Subsequent technical manuals have been produced through SCS/NRCS service centers since that time.

The Federal Highway Administration (FHWA) within the US Department of Transportation provides financial support and technical assistance for highway construction and maintenance throughout the US. These projects can have major environmental impacts upon the landscape. Based upon the work of Jones and Jones-landscape architects out of Seattle Washington [23] & [24]—the FHWA published their visual resource guidance document for highway projects in 1988 [25]. This document offers basic landscape and highway project visual inventory guidance as well as a visual impact methodology which can be used to compare transportation alternatives based on the attributes of landscape unity, vividness and integrity. The National Research Council-Transportation Board convened a panel in 2008 to reconsider the 1988 guidance for highway project visual impact assessment, and so hired a consultant group to undertake a study of VIA methods used by US State transportation agencies. See Palmer *et al.* [26], which summarizes part of this project as in regard to VIA practice across the US.

The US National Park Service (NPS) really has had no national guidance for visual or landscape management until recently. This is partially understandable as the NPS has, in many cases, separate legislation individually creating many of the national parks. One of the few comprehensive studies of landscape management alternatives was done for the Blue Ridge National Parkway in Virginia and Tennessee [27]. The author and James Palmer were involved in creating a landscape classification of the entire parkway corridor; developed color photographic simulations of different management alternatives (controlled burn vs. mechanical cutting of vegetation) and developed survey to test parkway user reactions to the simulated management alternatives. The NPS is in the process of developing multi-jurisdictional policy guidance for evaluating visual impacts of projects that may impact the scenic integrity of national parks. The NPS with US Environmental Protection Agency (USEPA) has sponsored extensive research on visibility degradation in the National Parks due to air pollution from 1979 [28] to the current time [29].

In 1988 the US Army Corps of Engineers (USACOE) published a report entitled *Visual Resources Assessment Procedure* (*VRAP*) for the US Army Corps of Engineers [30] & [31]. This was based upon the methodology development and training work by the author and James Palmer from1983 until 1988. It was also in response to a National Academy of Science panel that was convened to address why aesthetic and other environmental values were not properly being addressed by the USACOE as part of environmental assessment for water resource projects. This guidance was embraced as we developed better procedures to ensure reliability and validity as part of the VRAP methodology. As a result the VRAP methodology is fairly time consuming and costly, which explains why it has not been heavily utilized by the USACOE, but is has been used by other federal and state agencies for power plants and water related development.

A major conference, which coalesced much of the Federal agency activity was held at Incline Village, Lake Tahoe Nevada in 1979. This meeting was attended by over 800 people from 48 states and 11 countries with over 104 presentations and papers over three days [4]. It was co-sponsored by the US Forest Service, BLM, SCS/NRCS, FHWA and US Geological Survey and reflected the state-of-the art of visual landscape resource management and assessment at that time. Some of the best contributions from that conference were complied into the book *Foundations for Visual Project Analysis* [32] that becomes the first professional reference book for the field in North America.

2. Current VIA Challenges

After the 1980's there has been little new work in visual resources assessment methodology aside from developing visual simulation digital technology. This new technology and software is useful for improving visibility assessment, 3-D modeling and producing more realistic visual simulation of landscape change. Such examples can be seen in the book by Bishop and Lange [33] on *Visualization in Landscape and Environmental Planning* and a special issue of *Landscape and Urban Planning* [34]. There was also not much visual perception research as summarized in 1982 by Zube sell and Taylor's [35] landmark paper.

On the other side of the pond in Europe and particularly the United Kingdom (UK)—there have been developments in the standardization of the visual impact assessment process, which is summarized in the book *Guidelines for Landscape and Visual Impact Assessment* [36] produced by the Landscape Institute and the Institute of Environmental Management and Assessment. This book includes specific chapters on:

- Description of the proposed project;
- Baseline studies;
- Identification and assessment of landscape and visual effects;
- Presentation techniques & good practice;
- Consultation and review.

This basic VIA procedural guide has been utilized for large projects throughout the UK-particularly large wind farm projects. This brings us to the most recent issue affecting visual landscape assessment—renewable energy development. In the last 5 to 10 years we have seen a resurgence of VIA—particularly for large-scale wind turbine farms both on land and off shore. Some of these projects such as the proposed Cape Wind project between Cape Cod and Martha's Vineyard have proven to be very controversial [37] partially on aesthetic issues. These alternative energy development projects have caused the development of new VIA methods specifically for offshore wind farms in the UK [38]-[40], Australia [41] & [42] and in the US [43]-[47].

So in summary the early surge of visual impact assessment and visual resource management systems was spurred by NEPA in the late 1970's through the 1980's followed by stagnation in VIA/VRM method development. We had a steady development of digital visualization tools through the 1990's to the current time followed by recent VIA interest due to new renewable energy development including wind and solar energy farms, and large-scale biomass energy development.

Given the recent surge in renewable energy development—there has been a "slight surge" of VIA methodology development sponsored mainly by the US Bureau of Land Management and the National Park Service. Such work was highlighted in the 2012 National Association of Environmental Professionals (NAEP) meeting in Portland Oregon and includes:

- Visibility assessment of renewable energy projects [43]-[47];
- Air quality visibility impacts near and within national parks [29];
- Limited visual perception research in the US [20] [21] & [48] Germany [49] and Australia [42]; and
- New mitigation measure effectiveness [50].

The other notable research-to-practice development is the US Forest Services work on Appendix J [51] of the Scenery Management System. This is a deliberate attempt to clarify, refine and extend an ecological approach to scenery management. One of the key objectives of Appendix J is to provide a method for *sustaining valued scenery within an ecosystem management context*. The specific methodology includes:

- Using indicators to identify and measure ecological sustainability; and
- Using criteria for assessing six levels of visual disturbance, which indicates scenic integrity [52].

Although such method development toward a more ecological approach is in many ways desirable-it also presents a dilemma. The dilemma is whether to compare future visual/scenic landscape change against present conditions or a projected future incorporating ecological as well as human changes in landscape conditions. Clearly the later is preferred but we need a sound basis for making such projections. Then we can add in the future impact so climate change affecting ecological landscape conditions. Stephen Sheppard's new 2012 [53] book *Visualizing Climate Change* may give us some guidance in this regard.

3. Conclusion

The basic issue for this author is that the visualization technology has outstripped sound VIA methodology development. Such methodology needs to meet basic standards of validity, reliability and generalizability [16] [20] & [21]. This author has been an expert witness in court proceedings and quasi-judicial hearings over 35 years critiquing VIA methods used by consultants [54]. In many cases the basic step-by-step logic falls apart. Thus there is a need for best practice standards as well as sound VIA method development. Recently there has been progress both in the UK [38]-[40] and the US [44] & [55] with development of VIA methodology specifically for the assessment of large-scale wind, solar and geothermal energy development as well as mitigation measures [56] but there is much methodology research work to be done.

References

[1] Smardon, R.C. (1982) An Organizational Analysis of Federal Agency Visual Resource Management Systems. Unpublished Ph.D. Dissertation, Environmental Planning Program at the College of Environmental Design, University of California, Berkeley, 330 p. Available from International Microfilms, Ann Arbor.

[2] Smardon, R.C. (1986) Historical Evolution of Visual Resource Management within Three Federal Agencies. *Journal of Environmental Management*, **22**, 301-317.

[3] 42 USC s. 4321 et seq. The National Environmental Policy Act of 1969.

[4] Elsner, G.H. and Smardon, R.C. (Tech. Coord.) (1979) Proceedings of Our National Landscape: A Conference on Applied Techniques for Analysis and Management of the Visual Resource, Incline Village, Nevada, 23-25 April 1979, USDA For. Serv. Gen. Tech. Rpt. PSW-35, Berkeley CA, Pacific SW For. & Range Exp. Stn, 752 p. http://www.esf.edu/via http://www.treesearch.fs.fed.us/pubs/27530

[5] Bacon, W. (1979) The Visual Management System of the Forest Service, USDA. In: Elsner, G. and Smardon, R., (Tech. Coord.), Our National Landscape, 660. http://www.esf.edu/via http://www.treesearch.fs.fed.us/pubs/27530

[6] Ross Jr., R.W. (1979) The Bureau of Land Management and Visual Resource Management—An Overview. In: Elsner, G. and Smardon, R., (Tech. Coord.), Our National Landscape, 666-671. http://www.esf.edu/via http://www.treesearch.fs.fed.us/pubs/27530

[7] Schauman, S. and Adams, C. (1979) Soil Conservation Service Landscape Management. In: Elsner, G. and Smardon, R., (Tech. Coord.), Our National Landscape, 671-675. http://www.esf.edu/via http://www.treesearch.fs.fed.us/pubs/27530

[8] USDA, Forest Service (1973) National Forest Service Mgmt. Vol. 1, USDA Handbook 434, US Gov Print Off, Wash DC.

[9] USDA, Forest Service (1974) National Forest Landscape Mgmt. Vol. 2, Chap. 1. USDA Handbook 462, US Gov Print Off, Wash DC.

[10] Litton Jr., R.B. (1973) Landscape Control Points; a Procedure for Predicting and Monitoring Visual Impact. USDA, For. Serv. Res. Paper PSW-91, Berkeley, CA Pac. SW For. & Range Exp. Stn.

[11] USDA, Forest Service (1976) Forest Service Manual Sections 2360.01-2363.37 and 2350, US Gov Print Off, Wash DC.

[12] USDA, Forest Service (1995) Landscape Aesthetics; a Handbook for Scenery Management. Agricultural Handbook No. 701. US Gov Print Off, Wash DC. http://www.esf.edu/via http://www.fs.usda.gov/Internet/FSE_DOCUMENTS/fsm9_005639.pdf

[13] USDI, BLM (1980) BLM Manual Sections 6300, 6310 and 6320, Visual Resource Mgmt. USDI, Bureau of Land Management, Wash DC. http://www.esf.edu/via

[14] USDI, BLM (1980) Visual Resource Management, US Gov Print Off, Wash DC.

[15] USDI, BLM (1980) Visual Simulation Techniques, US Gov Print Off, Wash DC.

[16] Smardon, R.C., Feimer, N.R., Craik, K.H. and Sheppard, S.R.J. (1983) Assessing the Reliability, Validity and Generalizability of Observer Based Visual Impact Assessment Methods for the Western US. In: Rose, R.D. and Chestnut, L.G., Eds., *Managing Air Quality and Scenic Resources at National Parks and Wilderness Areas*, Westview Press, Boulder, 84-102.

[17] Feimer, N.R., Craik, K.H., Smardon, R.C. and Sheppard, S.R.J. (1979) Appraising the Validity of Visual Impact Assessment Methods. *Proceedings of Our National Landscape*, Incline

Village, 23-25 April 1979, 286-295. http://www.esf.edu/via
http://www.treesearch.fs.fed.us/pubs/27530

[18] Feimer, N.R., Smardon, R.C. and Craik, K.H. (1981) Evaluating the Effectiveness of Observer Based Visual Resource and Impact Assessment Methods. *Landscape Research*, **6**, 112-116. http://dx.doi.org/10.1080/01426398108705963

[19] Miller, P.A. (1984) A Comparative Study of the BLM Scenic Quality Rating Procedure and Landscape Preference Dimensions. *Landscape Journal*, **3**, 123-135.

[20] Palmer, J.F. (2000) Reliability on Rating Visible Landscape Qualities. *Landscape Journal*, **19**, 166-178.

[21] Palmer, J.F. and Hoffman, R.E. (2001) Rating Reliability and Representation Validity in Scenic Landscape Assessments. *Landscape and Urban Planning*, **54**, 149-161.
http://dx.doi.org/10.1016/S0169-2046(01)00133-5

[22] USDA Soil Conservation Service (1978) Procedures to Establish Priorities in Landscape Architecture. USDA SCS Technical Release No. 65, US Government Printing Office, Washington DC.

[23] Blair, G.E., Isaacson, L. and Jones, G.R. (1979) A Comprehensive Approach to Visual Resource Management for Highway Agencies. In: Elsner, G. and Smardon, R., Eds., *Proceedings of Our National Landscape*, Incline Village, 23-25 April 1979, 660.
http://www.esf.edu/via
http://www.treesearch.fs.fed.us/pubs/27530

[24] Jones G., Sorey, D. and Scott, C. (2012) Visual Resource Assessment for Regional Landscapes: A Modern Response to NEPA. 37*th Annual NAEP Conference on Science, Politics and Policy: Environmental Nexus*, National Association of Environmental Professionals, Portland Oregon, 21-24 May 2012, 641-656.

[25] USDOT and FHWA (1988) Visual Impact Assessment for Highway Projects. Publication No. FHWA-HI-88-054, USDOT, Federal Highway Administration, Washington DC.
http://contextsensitivesolutions.org/content/reading/visual-impact-2/

[26] Palmer, J.F., Churchward, C., Nassauer, J.I. and Swanwick, C.A. (2012) An Evaluation of the Visual Impact Assessment Process as Used by State Departments of Transportation in the USA. 37*th Annual NAEP Conference on Science, Politics and Policy: Environmental Nexus*, National Association of Environmental Professionals, Portland Oregon, 21-24 May 2012, 735-750.

[27] Noe, F.P. and Hammitt, W.E. (1988) Visual Preferences of Travelers along the Blue Ridge Parkway. Scientific Monograph Series Number 18, USDI, National Park Service Monograph Series No. 18, USDI, National Park Service, Washington DC.

[28] US EPA (1979) Protecting Visibility: An EPA Report to Congress. US Environmental Protection Agency, Office of Air Noise and Radiation, Office of Air Quality Planning and Standards; Research Triangle Park, North Carolina, EPA-450/5-79-008.

[29] Molenar, J.V., Paterson, R.W. and Goodstein, M.D. (2102) Seeking Clarity: Scenic, Economic and Legal Views on Visibility. 37*th Annual NAEP Conference on Science, Politics and Policy: Environmental Nexus*, National Association of Environmental Professionals, Portland Oregon, 21-24 May 2012, 767-817.

[30] Henderson, J.E., Smardon, R.C. and Palmer, J.F. (1988) US Army Corps of Engineers Visual Resources Assessment Procedure. *Transportation Research Record*, 67-71.

[31] Smardon, R.C., Palmer, J.F., Knopf, J., Grinde, K., Henderson, J.E. and Peyman-Dave, L.D. (1988) Visual Resource Assessment Procedure for US Army Corps of Engineers, Instruction Report EL-88-1, USACOE Waterways Experiment Station, Vicksburg, Miss.

http://www.esf.edu/via

[32] Smardon, R.C., Palmer, J.F. and Felleman, J.F. (1986) Foundations for Visual Project Analysis. John Wiley and Sons, New York. http://www.esf.edu/via

[33] Bishop, I.D. and Lange, E. (Eds.) (2005) Visualization in Landscape and Environmental Planning. Taylor and Francis, London.

[34] Lange, E. and Bishop, I. (Eds.) (2001) Our Visual Landscape; Analysis, Modeling, Visualization and Protection. *Landscape and Urban Planning*, **54**, 1-3. http://dx.doi.org/10.1016/S0169-2046(01)00121-9

[35] Zube, E.H., Sell, J.L. and Taylor, L.G. (1982) Landscape Perception: Research, Application and Theory. *Landscape and Urban Planning*, **9**, 1-33. http://dx.doi.org/10.1016/0304-3924(82)90009-0

[36] Landscape Institute/Institute of Environmental Management and Assessment (LI/IEMA) (2002) Guidelines for Landscape and Visual Impact Assessment. SPON Press, London.

[37] Williams, W. and Whitcomb, R. (2007) Cape Wind: Money, Celebrity, Class Politics and the Battle for Our Energy Future. Public Affairs Book, New York.

[38] DTI (2005) Guidance on the Assessment of the Impact of Offshore Wind Farms: Seascape and Visual Impact. Report of the Department of Trade and Industry Publication 806666 UK.
http://www.decc.gov.uk/assets/decc/what%20we%20do/uk%20energy%20supply/energy%20mix/renewable%20%energy/policy/offshore/rag/complete/file22852.pdf

[39] Hill, M., Briggs, J., Minto, P., Bagnall, D., Foley, K. and Williams, A. (2001) Guide to Best Practice in Seascape Assessment, Countryside Commission of Wales. Brady Simpson Marsh and University College Dublin.
http://www.marine.ie/NR/rdonlyres/683C8CDo-3367-47048542D3091607C9C2/0/interreg5_1.pdf

[40] Honor-McLennon and Envision (2006) Visual Representation of Wind farms: Good Practice Guidance. Report for the Scottish National Heritage, the South Scottish Forum and the Scottish Society for Direction and Planning. http://www.snh.gov.uk/docs/A305436.pdf

[41] Bishop, I.D. (2002) Determination of Thresholds of Visual Impact: The Case of Wind Turbines. *Environment and Planning B: Planning and Design*, **29**, 707-718. http://dx.doi.org/10.1068/b12854

[42] Bishop, I.D. and Miller, D.R. (2007) Visual Assessment of Offshore Wind Turbines: The Influence of Distance, Contrast, Movement and Social Variables. *Renewable Energy*, **32**, 814-831. http://dx.doi.org/10.1016/j.renene.2006.03.009

[43] Palmer, J.F. (2012) Maine's Experience Evaluating When Scenic Impacts from Wind Energy Development Are Reasonably Adverse. 37*th Annual NAEP Conference on Science, Politics and Policy: Environmental Nexus*, National Association of Environmental Professionals, Portland Oregon, 21-24 May 2012, 602-620

[44] Rooney, S., Dascomb, P., Korjeff, S. and Smardon, R. (2012) Technical Bulletin 12-001, Visual Impact Assessment Methodology for Offshore Wind Energy Conversion Facilities. Cape Cod Commission, Barnstable. http://www.esf.edu/via

[45] Sullivan, R.G., Kirchler, L.B., Cothren, J. and Winters, S.L. (2012) Preliminary Assessment of Offshore Wind Turbine Visibility and Visual Impact Threshold Distances. 37*th Annual NAEP Conference on Science, Politics and Policy: Environmental Nexus*, National Association of Environmental Professionals, Portland Oregon, 21-24 May 2012, 943-969.

[46] Sullivan, R.G., Cothren, J., Williamson, M., Smith, P., McCarthy, J. and Kirchler, L.B. (2102) Visual Impact Risk Assessment and Mitigation Mapping System for Utility-Scale

Wind Energy Facilities. 37*th Annual NAEP Conference on Science, Politics and Policy: Environmental Nexus*, National Association of Environmental Professionals, Portland Oregon, 21-24 May 2012, 667-692.

[47] Sullivan, R.G., Kirchler, L.B., Lahti, T., Roche, S., Beckman, K., Cantwell, B. and Richmond, P. (2012) Wind Turbine Visibility and Visual Impact Distances in Western Landscapes. 37*th Annual NAEP Conference on Science, Politics and Policy: Environmental Nexus*, National Association of Environmental Professionals, Portland Oregon, 21-24 May 2012, 897-942.

[48] Ribe, R. (2012) How Do People's Demographic Attributes and Environmental Attitudes Relate to Their Perception of Scenic Beauty? The Case of Forests in the Pacific Northwest. 37*th Annual NAEP Conference on Science, Politics and Policy: Environmental Nexus*, National Association of Environmental Professionals, Portland Oregon, 21-24 May 2012, 971-1025.

[49] Roth, M. (2012) Visual Landscape Quality Assessment in Landscape Planning in Germany: A Validation Approach. 37*th Annual NAEP Conference on Science, Politics and Policy: Environmental Nexus*, National Association of Environmental Professionals, Portland Oregon, 21-24 May 2012, 632-640.

[50] Schwarzler, K. and Cramer, G. (2012) The Art and Science of Camouflage for Visual Mitigation in the Landscape. 37*th Annual NAEP Conference on Science, Politics and Policy: Environmental Nexus*, National Association of Environmental Professionals, Portland Oregon, 21-24 May 2012, 693-715.

[51] USDA Forest Service (Undated) Briefing Paper, Scenery Management System-Appendix J. US Forest Service, Washington DC.

[52] Spencer, J., Mosier, J. and Laughlin, N. (2008) SMS Appendix J Project Level Scenery Analysis Template. US Forest Service.

[53] Sheppard, S.R.J. (2012) Visualizing Climate Change: A Guide to Visual Communication of Climate Change and Developing Local Solutions. Rutledge Press, London.

[54] Smardon, R.C. and Karp, J. (1992) The Legal Landscape, Guidelines for Regulating Aesthetic and Environmental Quality. Van Nostrand Rhinehold, New York.
http://www.esf.edu/via

[55] Sullivan, R.G. and Meyer, M. (2014) Guide to Evaluating Visual Impact Assessments for Renewable Energy Projects. Natural Resource Report NPS/ARD/NRR-2014/836 US National Park Service and Argonne National Laboratory.

[56] USDI Bureau of Land Management (2013) Best Management Practices for Reducing Visual Impacts of Renewable Energy Facilities on BLM-Administered Lands. USDI BLM, Wyoming Office, Cheyenne.

13

An Integrated Approach to Evaluate Benefits and Costs of Wastewater and Solid Waste Management to Improve the Living Environment: The Citarum River in West Java, Indonesia

S. M. Kerstens[1], G. Hutton[2], I. Firmansyah[3], I. Leusbrock[4], G. Zeeman[3]

[1]Royal HaskoningDHV, Amersfoort, The Netherlands
[2]UNICEF, New York, USA
[3]Sub-Department of Environmental Technology, Wageningen University, Wageningen, The Netherlands
[4]AEE INTEC, Gleisdorf, Austria
Email: sjoerd.kerstens@rhdhv.com, ghutton@unicef.org, indra.firmansyah@wur.nl, i.leusbrock@aee.at, grietje.zeeman@wur.nl

Abstract

Absence of wastewater and solid waste facilities impacts the quality of life of many people in developing countries. Implementation of these facilities will benefit public health, water quality, livelihoods and property value. Additional benefits may result from the potential recovery of valuable resources from wastewater and solid waste, such as compost, energy, phosphorus, plastics and paper. Improving water quality through implementation of wastewater and solid waste interventions requires, among others, an analysis of i) sources of pollution, ii) mitigating measures and resource recovery potentials and their effect on water quality and health, and iii) benefits and costs of interventions. We present an integrated approach to evaluate costs and benefits of domestic and industrial wastewater and solid waste interventions. To support a policy maker in formulating a cost- and environmentally effective approach, we quantified the impact of these interventions on 1) water quality improvement, 2) resource recovery potential, and 3) monetized benefits versus costs. The integration of technical, hydrological, agronomical and socio-economic elements to derive these three tangible outputs in a joint approach is a novelty. The approach is demonstrated using the heavily polluted Indonesian Upper Citarum River in the Bandung region. Domestic interventions, applying simple (anaerobic filter) technologies, were economically most attractive with a benefit cost ratio (BCR) of 3.2, but could not reach target water quality standards. To approach the target water quality, both advanced domestic (nutrient removal systems) and industrial wastewater treatment interventions were required, leading to a BCR of 2. We showed that benefits from selling recovered resources represent here an additional driver for improving water quality and outweigh the additional costs for resource recovery facilities. While included benefits captured some of the major items, these may have been undervalued. Based on these findings, water quality interventions justify their costs and are socially and economically beneficial.

Keywords

Sanitation, Water Quality Modeling, Economic Cost Benefit Analysis, Resource Recovery, Asia, Indonesia

1. Introduction

Nearly 40% of the population in developing countries lacks access to improved sanitation facilities [1], while an estimated 90% of all wastewater in developing countries is discharged untreated directly into rivers, lakes or the oceans [2]. Although access to improved sanitation facilities in South East Asia has reached 72%, Indonesia is lagging behind with only 61% [1]. Moreover, Indonesia, like other developing countries, largely lacks solid waste management services and suffers from uncontrolled discharge of industrial wastewater [3] [4]. The absence of domestic and industrial wastewater and solid waste facilities is associated with a number of impacts.

First, discharge of untreated sewage can lead to adverse health effects on individuals [5]. Health conditions can be improved by wastewater and hygiene interventions [6] [7]. Reference [8] showed that *E. coli* concentrations in canals could be substantially reduced (~4 log) by sewage collection and treatment. On-site sanitation (e.g. pit latrines as commonly applied in Indonesia) in combination with shallow ground water sources and high population density may also impact public health [9] [10].

Second, discharge of untreated wastewater increases nitrogen (N), phosphorus (P) and organic pollutants (Chemical oxygen demand (COD) and Biological Oxygen Demand (BOD)) loads to water bodies. This may result in eutrophication and low oxygen levels in waters, thus impacting ecosystem functioning [11]-[14]. Domestic pollution depends on living conditions and type of residential areas [15]-[17]. There is also a positive correlation between imperviousness and urban density on pollution gradients in receiving water bodies [18]. A range of wastewater and solid waste systems exists and their feasibility can be linked to residential features [19] (see Online Supplementary Information (OSI), Section 1). Wastewater and solid waste systems range from conventional (e.g. solid waste landfilling) to those that reduce, reuse and recycle (3R) solid waste components or recover resources from wastewater [19] [20]. Industrial wastewater discharge may also contribute significantly to water pollution [14]. Applicable treatment technologies depend on type of industry, biodegradability, toxicity, robustness, effluent standards or reuse requirements [21] [22]. Water quality is further affected by agricultural activities, as a result of fertilizer use, aquaculture and livestock emissions [11] [23].

Third, the value of recoverable resources from wastewater and solid waste, such as energy, water, organics, nutrients, plastic and paper is frequently neglected, whereas the sale of recovered resources can assure long-term operational and financial sustainability [24]-[28]. The potential demand for recovered resources depends on agricultural activities and possibilities to replace conventional production processes using virgin materials by processes using recyclables (paper and plastics) [29].

Finally, the absence of wastewater and solid waste facilities may accrue socio-economic impacts, such as travel and waiting time for community or public toilet facilities, loss of social capital and equity and decreased property values [14] [30]-[33].

Thus, implementation of wastewater and solid waste interventions benefits public health, the environment, resource conservation, the economy and people's welfare. However, given that implementation of interventions involves costs in the form of investments, operation and maintenance of the facilities, policy makers need to understand the outcomes (benefits) of major actions in relation to their costs [34]. The Benefit Cost Ratio (BCR) describes benefits of intervention relative to its costs. Given that benefits may require a long time to manifest and planned infrastructure are designed for long lifetimes, benefit-cost analysis should use a time horizon of at least 20 years [35]. A demonstrated BCR of one or more—indicating a return on investment of at least 1.0 given the discount rate used—can feed into advocacy efforts to raise funding from governments and households, and can convince the private sector to invest [32].

Individual cause-effect relationships to evaluate the costs and benefits to improve water quality have been established, such as: 1) the effect of pollution load on the quality of receiving water [13] [18], 2) the effect of sanitation on public health improvement and reduced discharged pollution loads [11] [36], 3) economic losses as a result of poor sanitation [32], and 4) technical and financial feasibility of wastewater and solid waste technologies [19]. However, no integrated framework exists in the scientific literature that quantifies the effect of applicable wastewater and solid waste interventions on 1) water quality, 2) resource recovery potential, and 3) monetized benefits and costs. This paper therefore proposes to use a combination of methods that describe these individual cause-effect relationships, and synthesize them to produce these three tangible outputs. This multi-methods approach allows policy makers to make well-informed choices in wastewater and solid waste planning.

The developed approach can be used on any river basin or delta. In this paper, the Upper Citarum River in West Java (Indonesia) is used as a case study because of its very low water quality combined with its impact on the life of millions of people downstream (see **Figure 1** and OSI Section 2).

2. Materials and Methods

To assess the impact of wastewater and solid waste interventions on water quality and estimate resource recovery and economic returns, the following six consecutive steps were formulated (**Figure 2**). In step 1 the river water quality at different locations was

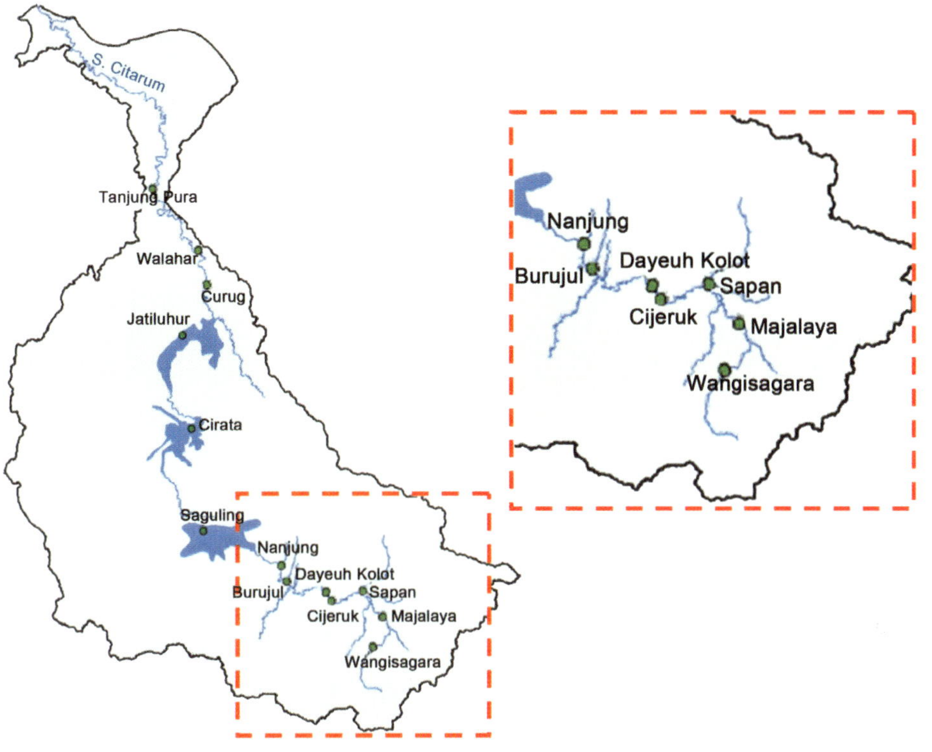

Figure 1. Location of the Upper Citarum River basin (in box) within the Citarum basin.

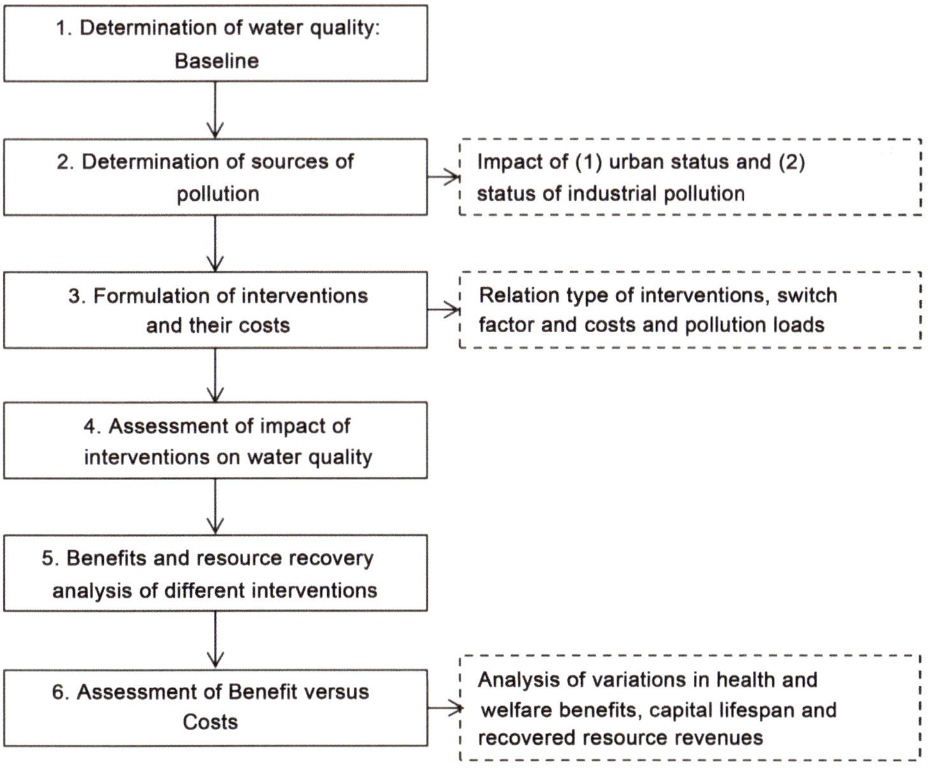

Figure 2. Approach applied to determine the BCR of interventions (Dashed blocks show activities for which a sensitivity analysis was performed).

collected. This information was used as a baseline to determine the impact of different types of interventions. In step 2 the sources of pollution COD, BOD, N and P per sector (domestic, industrial and agricultural) were determined. An additional assessment on the relative contribution per sector was performed considering variations in the pollution load reaching the surface water with different urban areas [18] and the status of industrial pollution control [37]. In step 3 wastewater and solid waste interventions were defined and their associated costs estimated, based on [19]. The impact on pollution loads discharged to the environment and the associated costs were further analyzed by varying treatment technologies and the percent of households switching from a septic tank to a sewer system connection. In step 4 the impact of different interventions on water quality was determined using a river basin simulation software (RIBASIM) [38]. In step 5, five different benefits were monetized: health, access time, water quality, environment and revenues from resource recovery [19] [31]. In step 6 the benefits and costs were compared over a 20 year period to estimate the benefit-cost ratios. In this final step also a sensitivity analysis was performed to determine the impact of reduced health, welfare and revenues from recovered resources and of different capital lifespan on the BCR. A description of the individual steps and method for data collection is further illustrated using the Upper Citarum River as an example.

2.1. Step 1: Determination of Water Quality in Upper Citarum River

Water quality data for COD, BOD, N and P for the period 2001-2009 in the upper Citarum River at Wangisagara, Sapan, Cijeruk, Dayeukholot and Nanjung (**Figure 1**) was obtained through the West Java Regional Environmental Agency [39] [40].

2.2. Step 2: Determination of Sources of Pollution

Three sources of pollutions were distinguished and assessed for 2010 and 2030, being (A) Domestic, (B) Industrial and (C) Agricultural (**Table 1**).

A. Domestic pollution:

Domestic pollution was determined in five steps.

1. Determination of specific per person pollution loads: Domestic specific water consumption rates followed the Indonesia guidelines [16] for 6 categories of urban area: 1) metropolitan (>1 million people), 2) large town (500,000 - 1 million people), 3) medium town (100,000 - 500,000 people), 4) small town (20,000 - 100,000 people), 5) village (3000 - 20,000 people) and 6) rural (<3000 people). An 80% return factor was used to estimate wastewater production from consumed water [41]. Metropolitan specific pollution loads were based on [19].
2. Correction of pollution load with varying types of urban status: The relation between urban category and pollution loads was reflected using the study of [17] applying a greywater pollution load decrease between urban metropolitan and rural areas of 30% of COD and N and 50% of P, while for urban categories in between metropolitan and rural areas these were made relative to water consumption data (**Table 1**). Because of lack of detailed data, urban and rural black water pollution load rates were assumed the same.

Table 1. Basis for applied Domestic (A), Industrial (B) and Agricultural (C) pollution reaching the surface water.

A. Domestic per capita pollution loads reaching surface water[a]							B. Industrial concentrations in effluent per type of industry				
Urban category	Water use	COD	BOD	TN	TP	Coliform	Type of industry	COD	BOD	TN	TP
	l/cap/d	g/p/d				1/100 ml		mg/l			
1. Metropolitan	190	82.2	41.1	12.3	2.1	1×10^8	Food & Beverage[b]	5000	3000	80	30
2. Large town	170	81.0	40.5	12.3	2.0	1×10^8	Paper[c]	4000	1500	20	10
3. Medium town	150	73.5	36.7	11.3	1.9	1×10^8	Pharmaceutical[d]	5000	1500	127	25
4. Small town	130	65.3	32.7	10.2	1.7	1×10^8	Rubber[d]	7340	4400	1100	220
5. Village	100	56.9	28.5	9.1	1.5	1×10^8	Textile[e]	1350	450	60	20
6. Rural	30	47.3	23.7	7.9	1.3	1×10^8	Others[d]	280	168	42	8

C. Agricultural pollution loads (g/Yield.ha)[d]

Type of crops	COD	BOD	TN	TP	Coliforms
Rice	45	22.5	21.5	6.5	0
Non-rice food crops	34	17	4.6	0	0

a. Based on [17]-[19]; b. Data obtained by authors from Food & Beverage (dairy, brewery) in Indonesia; c. Values depend on type of paper and pulping process and range from 1500 to over 20,000 mg/l COD [21] [48]. Applied values are based on experience of authors for Pulp and Paper South East Asia; d. Based on [47]; e. Textile industry data were determined based on actual measurements of 21 textile industries in project area [37] and verified with [21].

3. Correction of pollution reaching surface water bodies: Baseline pollution correction coefficients (included in **Table 1**) were based on [18] and were 100% (metropolitan and large towns), 92% (medium town), 83% (small town), 74% (village) and 65% (rural areas). Thus, only 83% of pollution generated in a small town is expected to reach the surface water. As specific information on these coefficients was lacking for the Upper Citarum Basin, two alternative scenarios were compared, being 1) where 100% of pollution entered the surface water, and 2) where half of the baseline value entered the surface water (i.e. 50% for metropolitan and large towns, 46% for medium town, 42% for small towns, 37% for village and 33% for rural areas).
4. Determination of pollution loads reaching the surface water for 2010 and 2030: Total specific pollutions loads per location reaching the surface water were calculated applying the specific pollution loads (combining step 2 and 3 above) on population developments obtained from the Java Spatial Model (JSM). JSM shows the population development for each urban category between 2010 and 2030 [16].
5. Determination of the number of people with access to wastewater facilities in 2010: The pollution loads reaching the water bodies were corrected for interventions already in place. The 2010 wastewater access data were obtained from the statistical bureau of Indonesia (BPS) and were determined as 52%. 490,000 people were connected to the Bojong Soang WWTP (pond systems) in Bandung [42].

B. Industrial pollution:

838 industries in the catchment area were categorized by location and type (**Table 1**) and water consumption (m^3/d) in which data on ground and surface water consumption were obtained from the West Java provincial agency for Energy and Mineral Resources [43] and provincial agency for Water Resources Management [44]. Pollution loads were determined by effluent flow (using 80% return factor) and effluent concentrations (**Table 1**). Because reliable industrial pollution data is lacking [37], an impact analysis was performed (**Table 2**). A distinction is made between 1) a best case scenario, 2) a baseline scenario and 3) a worst case scenario. These scenarios vary in terms of removal efficiency and percentage of industry having a WWTP (**Table 2**). COD removal efficiencies in the best case followed self-reported COD removal efficiencies by industries, whereas the worst case effluent COD values followed externally measured COD removal efficiencies [37]. N and P are not measured by industries and presented values were assumed, based on [21]. Reference [45] reports that 80% of the textile industries comply with the effluent standards, whereas the environmental office in nearby Cimahi mentions 3% [46]. Therefore, the baseline case assumes that 80% of the largest industries (consumption >2000 m^3/d) treat their wastewater, while with decreasing water consumption this percentage decreases with a minimum of 25% (**Table 2**).

C. Agricultural pollution:

The 2010 and 2030 water demand for irrigation was based on [16]. Pollution discharged (**Table 1**) for rice and non-rice crops were based on [47].

2.3. Step 3: Formulation of Interventions and Their Costs

Domestic interventions:

Selection of type of domestic WWT facilities (**Table 3**) was based on the residential features following [19]. For off-site systems three scenarios were compared to identify the effect on the surface water quality and cost:

1. Simple Technology (ST): Anaerobic filter is applied for medium centralized systems and a conventional activated sludge (CAS) for centralized systems;
2. Advanced Technology (AT): Medium central and central systems apply a CAS with additional N, P removal;

Table 2. Defined scenarios to determine the impact of industrial pollution loads by varying 1) removal efficiencies and 2) availability of WWTP based on size of water intake[a].

Scenario	1. % Removal efficiency				2. % industries with WWTP per size of water intake (m^3/d)				
	COD	BOD	TN	TP	0 - 100	100 - 500	500 - 1000	1000 - 2000	>2000
Best case	90	95	90	50	35	35	60	80	90
Baseline	65	69	65	36	25	25	50	70	80
Worst case	40	42	40	22	15	15	40	60	70

a. **Table 2** shows for example that in the best case scenario, 90% of the industries with a water consumption exceeding 2000 m^3/d have a WWTP and removal efficiencies are 90% (COD), 95% (BOD), 90% (TN) and (50% (TP).

Table 3. WWT system selection based on 1) population density and 2) urban/rural category. Removal efficiencies of Simple technologies (ST), Advanced Technologies (AT) and Resource Recovery (RR) technologies for COD, BOD, TN, TP and coliforms are based on [19].

System	Criteria for use[a]		Applied removal efficiencies per type of technology									
	Residential population density	Status 2020[b]	COD (%)		BOD (%)		TN (%)		TP (%)		Coliforms (%)	
			ST	AT/RR	ST	AT/RR	ST	AT/RR	ST	AT/RR	ST	AT/RR
On-site	<100	Rural/Urban	40[a]		45		15		5		90	
CBS	>100	Rural	80[a]		85		15		5		99	
Medium Central	100 - 250	Urban	80	88	85	97	15	90	5	67	99	99.9
Central	>250	Urban	88		97		73		29		99.9	

a. Current users in urban areas with a residential density between 25 - 100 pp/ha apply on-site systems, whereas all new development will be served by medium centralized system [50]; b. Selection criteria are formulated based on the expected population status in 2020 (mid-term).

3. Resource Recovery technology (RR): Comprising Aerobic Granular Sludge (AGS) system [49] with sludge digestion, P-recovery as struvite and composting of produced sludge. The removal efficiencies of AT and RR are the same.

Associated investment and operational costs were based on [19] (see Table S1 of the OSI Section 1). The effects on discharged pollution loads reaching the surface water and associated investment costs of a 25%, 50% and 75% switch of households currently applying on-sites system to an off-site system were compared.

Industrial interventions:

Three industrial wastewater treatment types were formulated based on currently applied technologies [51] (see OSI, Section 3): 1) textile wastewater using reactive dyes (typically used for traditional batik), apply a CAS and activated carbon for color removal, 2) textile wastewater using non-reactive dyes apply CAS followed by Dissolved Air Flotation (DAF), and 3) other industries apply pre-treatment (DAF) and CAS. Future effluent values should meet at least current standards [46] defined as 80 mg/l COD, 20 mg/l BOD, 10 mg/l N and 10 mg/l P. Investment and operational costs were determined for different sizes of treatment capacities, based on available engineering cost standards (see OSI, Section 3).

Municipal Solid Waste (MSW) interventions:

Solid waste system selection interventions (**Table 4**) and their costs are based on [19] and distinguish home composting, landfilling and centralized and decentralized 3R application (see OSI Section 4).

2.4. Step 4: Assessment of Impact of Interventions on Pollution Loads and Water Quality

A generic model package (RIBASIM) for simulating the behavior of river basins under various hydrological conditions was used to simulate the effect of different interventions on water quality development in the Upper Citarum River [38] [52]. Based on pollution loads produced in each defined catchment area and resulting water flows

Table 4. MSW system selection for Indonesia as a function of density, urban/rural status [50].

Type of area & Density / Activity	Rural		Urban	
	<25 pp/ha	>25 pp/ha	<100 pp/ha	>100 pp/ha
Collection	no	yes	yes	
Disposal	no	yes	yes	
Level of 3R	Home composting	Decentralized composting and plastic/paper recovery	Central digestion and composting and plastic & paper recovery	

concentrations are calculated. The RIBASIM model and defined catchment areas are further explained in OSI Section 5. The pollution loads entering the Upper Citarum River were varied, using 6 scenarios (**Table 5**).

The output of the 2010 RISBASIM average pollutant concentrations was calibrated based on the average measured concentration (step 1).

2.5. Step 5: Benefits Analysis of Different Interventions

Five economic benefits of wastewater and solid waste management improvements were defined following [32] and [53]:

A. Health:

Averted costs of fecal-oral disease from improved on-site sanitation and wastewater management: An average disease reduction of 36% by on-site sanitation and an additional 20% by adding improved off-site facilities was applied [54]-[56]. The average annual health cost per 5 member family as a result of unimproved sanitation was US $316 [31].

Associated averted health impacts (infectious diseases and skin complaints) of less exposure during flooding events: Reported health cases during a period of several flooding events (January-March 2009) were compared to the same period in a non-flood year (January to March 2010) and was scaled to reflect all the flooded communities in the Citarum River basin, resulting in an estimated 15,000 averted cases of diarrhea in an average year [53]. The economic value was estimated by multiplying the average number of additional cases per year by the unit cost of inpatient (hospitalized) and outpatient services, including productivity losses [31].

B. Access time:

Value of time savings from reduced travel time and/or queuing for meeting sanitation needs: An average daily gain of 115 minutes per household with an annual value of US $95 per household is used [31]. Only the time of adults and school-aged children were included, valued at 30% and 15% of the hourly rate implied by the GDP per capita, respectively [57]. This figure was applied to the access gain afforded by on-site sanitation facilities of 45% of households for the period from 2010 until 2030.

C. Water:

Reduced drinking water treatment costs to households and industries: The total cost of water treatment (including both capital and operating costs) using surface water of a better quality source will decrease from 0.13 to 0.06 US $/m^3 [16]. This saving was

Table 5. Defined intervention scenario (S1 - S6); ST = Simple Technology ; AT = Advanced Technology and percentage of population served by a municipal solid waste (MSW) system.

Name		Description
S1: Baseline		2010: Baseline situation
S2: No intervention		2030: Baseline case; same WWT access percentage as 2010 applied. Only correction for population growth for WWT and MSW
S3:	25% ST	2030: 100% Domestic access, use ST and 25% switch + 100% MSW
	25% AT	2030: 100% Domestic access, use AT and 25% switch + 100% MSW
	50% ST	2030: 100% Domestic access, use ST and 50% switch + 100% MSW
	50% AT	2030: 100% Domestic access, use AT and 50% switch + 100% MSW
	75% ST	2030: 100% Domestic access, use ST and 75% switch + 100% MSW
	75% AT	2030: 100% Domestic access, use AT and 75% switch + 100% MSW
S4: Industrial only		2030: Industrial WWT intervention; 100% of big (>1000 m^3/d), 90% of medium (500 - 1000 m^3/d), 80% small (100 - 500 m^3/d), and 75% of very small (<100 m^3/d) sized industries apply intervention. Domestic WWT, MSW interventions follow S2
S5: 25% - 75% ST/AT		2030: Combination of scenario 3 and 4
S6a: 25% - 75% RR		2030: Same as S5, using recovery technologies for domestic, industrial effluent recycling and MSW

a: Except for S6, where a MSW resource recovery based system is applied, all other cases apply a conventional MSW system (no resource recovery).

multiplied by the assessed annual production of water from surface water sources (207 million m^3 for domestic and 70 million m^3 for industrial consumers) in 2030.

Improved fish yields from farming in downstream lakes due to improved water quality: Data collected through interviews with the regional Fisheries Office showed a decrease in fish catch of 5,000 ton/year in recent years [58]. Fish kills in Saguling (**Figure 1**) related to discharge of untreated wastewater have been described by [12] and [23]. By 2030 the fish capture is estimated to increase by 8,000 metric tons per year [58]. The increase of improved water quality was assumed to account for one-third of this expected annual gain of farmed fish in the Citarum basin [53]. A market prices of fish of 1.5 US $/kg was used [58].

D. Environment:

Reduced frequency of river and reservoir dredging due to improved sludge and waste management: An estimated 35 l/person/year of septic waste [9] and 11% and 17% of domestic urban and rural solid waste [59] accumulating to nearly 500 ktonne/year are currently discharged to the surface water and will be prevented from being disposed in the surface water in 2030 with the described interventions (**Table 5**). With a cost of dredging estimated at US $3.76 [16] per ton of sediment (assuming no degradation), the total annual cost averted was estimated.

Rise in land prices due to improved aesthetics of riverside and lakeside real estate: Currently the Citarum riverside area is not developed due to water pollution. However, the area is expected to become a place where riverside property could be developed for inhabitants, small businesses, and tourist facilities in a situation where water quality is improved. The current agricultural land price (10.7 US $/m^2) in the vicinity of Bandung was used as a benchmark for current riverside land prices. The current market suggests that land prices can climb to 71.3 US $/m^2 in highly desirable locations [16]. In this study 50% of this increase is attributed to improved water quality. This value was multiplied by an estimated 50 ha of land that could be developed each year after the water quality improvements have occurred [53].

Averted maintenance costs of hydro-electric facilities: Improved solid waste management would avert the current costs of US $0.1 million [16] to evacuate the solid and unmanaged sludge waste to avoid equipment damage in the hydroelectric facility [53].

E. Recovery of resources:

In scenario 6, resource recovery was considered (see also Table S4 in the OSI, Section 4):

- Off-site wastewater systems: Production of energy (sludge digestion), struvite (from centrate) and compost (digested sludge composting).
- MSW: Energy and compost production from organic waste and recovery of plastics and paper.
- Industrial wastewater: industries with a water consumption exceeding 2000 m^3/d reused 80% of the effluent, whereas for industries using 1000 - 2000 m^3/d this was 50%.

To compare the production (recovery) of resources with the potential demand in the Upper Citarum River catchment area in 2030, the compost, struvite, plastic and paper demand in the whole of West Java obtained from [29] was corrected for people living in the Upper Citarum River basin area. The amount of recycled water from industries was compared to the total domestic and industrial water demand in 2030 in the catchment area [16]. Energy production from digestion is compared to the energy demand for domestic wastewater treatment in the area applying aerobic granular sludge technology [19].

2.6. Step 6: Assessment of Benefits versus Costs

To relate benefits and costs to either wastewater or solid waste interventions, BCR's were presented separately. To analyze the individual impact of domestic, industrial and resource recovery interventions the BCR of scenarios S3: (50% ST and S3: 50% AT), S4: (Industrial interventions only), S5: (50% ST; S5: 50% AT) and S6 (50% RR) were determined (see also **Table 5**). A sensitivity analysis was performed in which input values that have the highest anticipated impact were varied: (1) health and access time benefits reduced from 100% to 50%, (2) lifespan of all wastewater and solid waste facilities varied from 20 years to 15 and 40 years, and (3) resource selling price reduced to half baseline values [32] [53]. Health and access time benefits were all attributed to domestic intervention. Water quality and environmental benefits were attributed to the fraction of COD load discharged by domestic and industrial sources respectively.

3. Results

3.1. Water Quality in Upper Citarum River

Figure 3 shows the average 2000-2009 water quality from upstream to downstream locations. Maximum allowable concentrations are defined in class II standard [60] and are COD 25 mg/l, BOD 3 mg/l and P 0.2 mg/l. From Sapan on (**Figure 1**) all measured values exceed these standards. Concentrations in several Citarum branches passing high density urban areas, show COD values approaching 500 mg/l and pathogen levels as high as 10^7 Units/100 ml [40].

3.2. Sources of Pollution

Current cumulative pollution loads in the Upper Citarum River basin of COD (585 tonne/d), BOD (264 tonne/d), TN (91 tonne/d) and TP (20 tonne/d) were determined as the baseline values (**Table 6**). The sensitivity analysis with variations in domestic pollution coefficient [18] and performance of industries shows considerable differences with the baseline scenario (**Table 6**) with COD loads varying between 325 and 688 tonne/d (see also OSI, Section 7).

3.3. Effect of Selected Interventions on Costs and Pollution Loads

The domestic pollution loads entering the Upper Citarum River depend on (1) the type of technology applied (simple versus advanced) and (2) the rate of current households applying on-site systems in urban areas that will switch to an off-site system (**Figure 4**). The use of advanced compared to simple technologies has a minor impact on COD removal in the range of 3% - 4%, but a major impact on N-removal in which a rate of

Table 6. COD, BOD, TN and TP pollution loads reaching the surface water by source for the baseline scenario and varying pollution correction factors and industrial practices.

Source	Scenario	COD (tonne/d)	BOD (tonne/d)	TN (tonne/d)	TP (tonne/d)
Domestic	Baseline loads	388	188	68	12
	100% reaches surface water	440	213	78	14
	Half of baseline loads reach surface water	194	94	34	6
Industrial	Baseline	163	60	6	2.6
	Best case	98	33	4	2.2
	Worst case	215	80	8	3.0
Agriculture		34	17	16	5
Total	Baseline[a] (S1)	585	264	91	20
	Minimum[b]	325	144	54	13
	Maximum[c]	688	310	103	22

a. Total baseline values comprise domestic and industrial baseline loads + agricultural loads; b. Total minimum values add domestic low pollution correction coefficient and Industrial best case + agricultural loads; c. Total maximum values add domestic high pollution correction coefficient and Industrial worst case + agricultural loads.

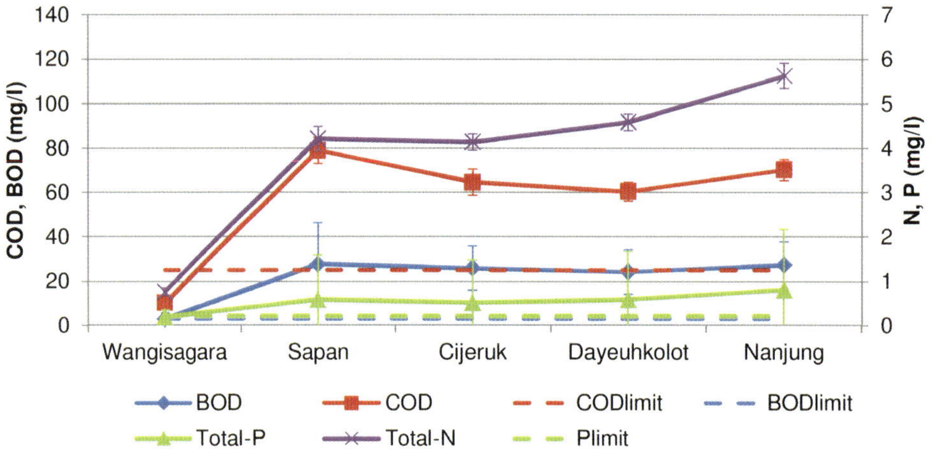

Figure 3. Average and standard variations of COD, BOD (primary y-axis) and N, P (secondary y-axis) concentrations at indicated locations in the upper Citarum (2000-2009) [39] and COD, BOD and N limits.

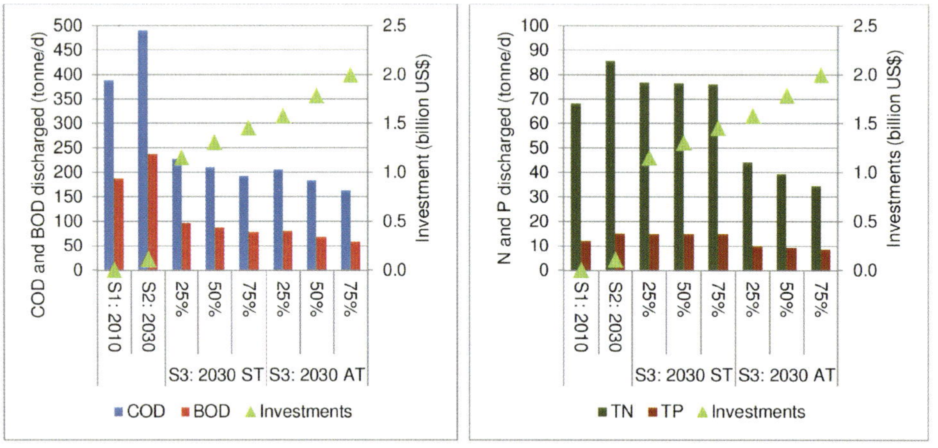

Figure 4. Calculated domestic COD, BOD (left) and N, P (right) pollution loads per type of intervention and their investment costs (secondary y-axis). S1 (baseline), S2 (no intervention) and S3 (domestic interventions) applying simple (ST) or advanced technologies (AT) with increasing (25%, 50% and 75%) values for urban on-site users that switch to off-site systems.

25% households switching to off-site systems leads to a 29% difference and a rate of 75% households switching to off-site systems leads to a 37% difference (**Figure 4**).

When increasing the switch factor from 25% to 75%, the additional removed COD and N increased with 5% and 1% for simple technologies and 6% and 9% for advanced technologies. BOD removal follows the COD trend, whereas P removal follows the N trend. Thus, the application of advanced technologies or a higher rate of people switching from on-site system to off-site systems mainly affects the additional nutrient removal, while organic removal is less affected. The numeric values of this analysis and further elaboration on costs of interventions and their impact on water quality are described in the OSI, Section 7.

The industrial pollution load amounts to 28% of the total load (**Table 6**), but industrial interventions can result in a relatively large COD reduction (35%) compared to the combined domestic and industrial COD reduction (see also OSI, Section 7).

3.4. Effect of Interventions on Water Quality

Figures 5(a)-(f) shows the effect of interventions on the year round average water quality at different locations. The location names are approximate locations, as RIBASIM calculates concentrations in defined segments of a river (see OSI, Figure S8). Without

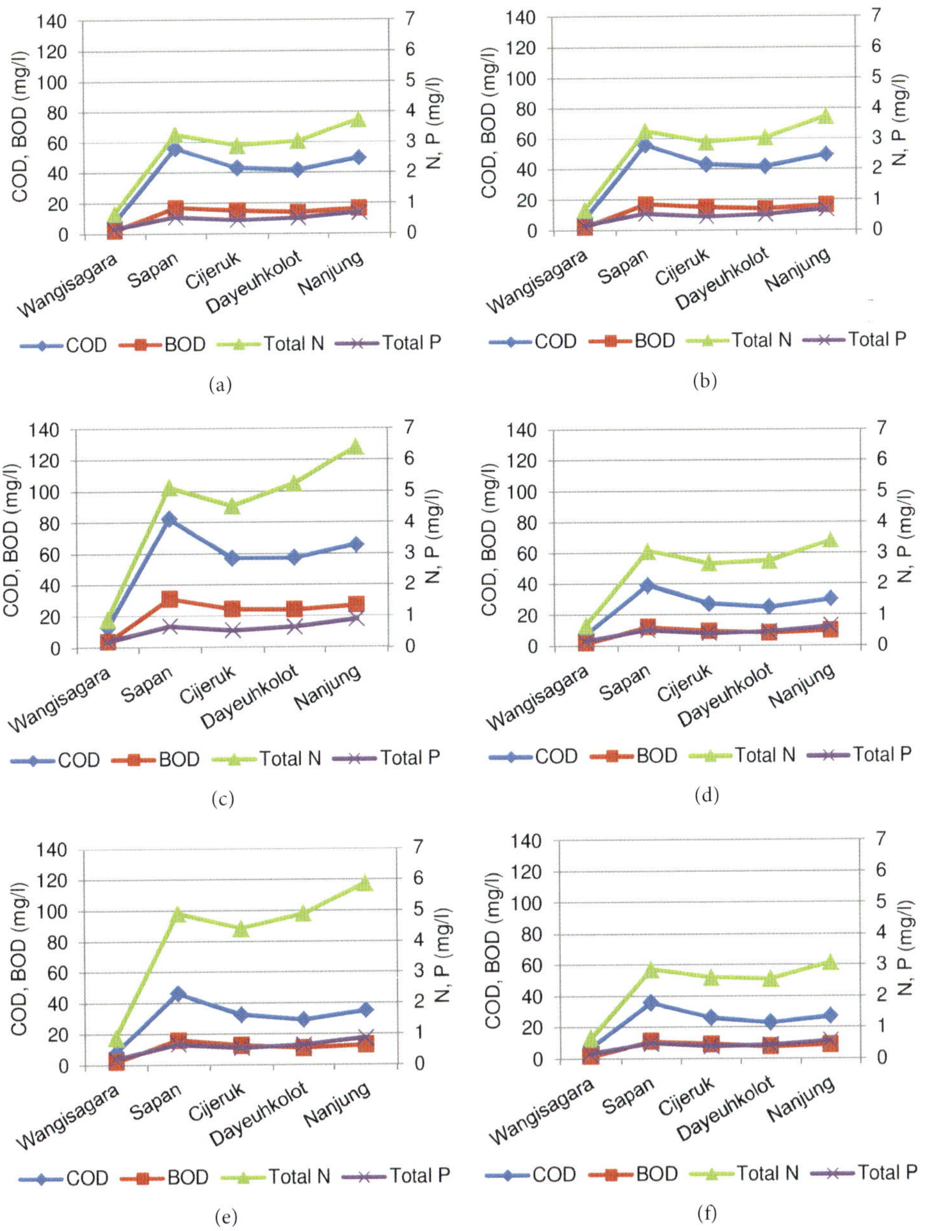

Figure 5. Modelled COD, BOD (primary y-axis) and N, P (secondary y-axis) concentrations at indicated locations in 2030 with varying switch factors % and simple (ST) or advanced technologies AT). (a) S2, no intervention; (b) S3: 50% AT); (c) S4: Industrial only; (d) S5: 50% AT; (e) S6: 25% ST; (f) S6: 75% AT. Limits for COD, BOD, and P are 100 mg/l, 3 mg/l and 0.2 mg/l.

additional interventions all concentrations will increase compared to the 2010 values (**Figure 3**) with values as high as 100 mg/l of COD (**Figure 5(a)**). The modeled pollutant concentrations in water entering Saguling reservoir (approximate location Nanjung) are 80 mg/l COD, and 7 mg/l TN and 1 mg/l TP. When applying S3 with 50% AT (**Figure 5(b)**) a considerable drop in all pollution concentrations is achieved, whereas the introduction of industrial interventions result in approximately 20% COD & BOD removal and about 4% N & P removal (**Figure 5(c)**). The combination of these interventions (S5: 50% AT; **Figure 5(d)**) results in concentrations of 30 mg/l for COD, 10 mg/l for BOD, 3.4 mg/l for TN and 0.7 mg/l for TP. The maximum removal scenario (**Figure 5(f)**) results in values approximating the class II standard (COD < 25, BOD < 3, P<0.2 mg/l). Comparing **Figure 5(e)** (ST) with **Figure 5(f)** (AT) shows limited impact on COD or BOD removal, while considerable extra N, P removal is shown when using advanced instead of simple domestic technologies.

To reach the desired water quality levels (class II) both industrial and domestic municipal interventions are needed. In addition, the applied off-site technologies should also include N and P removal, requiring more advanced and more costly technologies (see **Figure 4**) compared to the application of only anaerobic filters.

3.5. Benefits of Interventions

The maximum quantified economic benefits are US $430 million per year in which health benefits account for 39% (**Figure 6**). Health benefits largely result from reductions in fecal-oral diseases, since 1) the people without access to wastewater (on-site and Bojong Soang WWTP) facilities (48%) all have access by 2030 (55.2% of health benefits), and 2) people that have access to a well-managed off-site or fecal sludge management system increased from 7% to 73% (44.6% of health benefits). Associated averted health impact due to irregular flooding events is only US $0.3 million.

Convenience and time savings are among the top five reasons for having a latrine in the home area [31]. Based on [31] a mean annual gain of US $77 million was determined for an additional 45% of the population in 2030 having access to their own

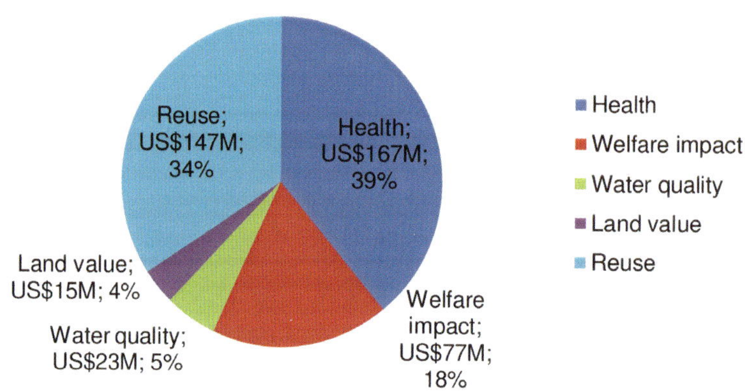

Figure 6. Contribution of calculated overall economic benefits expressed in million US $ (total US $430 million) of each monetized impact (Scenario 6). Sedimentation (US $2M; 0% contribution) and Dam maintenance (US $0.1; 0% contribution) are not shown.

latrine facilities. This estimate is conservative as 1) it excludes travel needs for urination purposes, and 2) time is valued conservatively at 30% of the GDP per capita at hourly values.

US $13.9 million of the total US $23 million reduction in water treatment cost will accrue to the public water utilities and their consumers, while industries are expected to benefit US $4.7 million annually. The value of farmed fish yields is expected to be US $4 million annually.

The combined environmental benefits (increased land value, reduced dredging, averted maintenance costs of hydro-electric facilities) amount to US $17 million, of which nearly 90% is attributed to increases in land value based on annual land sales. The benefits of reduced dredging (even assuming no decomposition or organic waste) have minor benefits.

Table 7 shows the estimated reuse benefits based on the per capita production features and resource values (Table S4 in OSI). 87% of the US $147 million yearly potential revenues are from municipal solid waste, 11% from domestic wastewater treatment and recovery and reuse of its resources and recycling, and 2% from industrial wastewater treatment and recycling. The potential demand for recoverable resources is higher than the potential supply through recovery (**Table 7**), ranging from a factor 13 for water to a factor 1.6 for plastic.

3.6. Assessment of Benefits versus Costs

Following the anticipated benefits (**Figure 6**) and corresponding investment and operational costs (Table S6 in OSI, Section 8) the BCR was calculated (**Figure 7**). The BCR varied between the interventions. The highest BCR of 3.2 is achieved by implementing simple technologies (S3: 50% ST), in other words an economic return of US $3.2 is anticipated for each US $1 invested. Because of higher costs for AT compared to ST, the BCR is expected to be lower for the AT scenario (BCR in S3: 50% AT = 2.06). The lowest BCR (0.52) is found in scenario 4 (industrial interventions alone). A joint approach

Table 7. Resource recovery potential, sector of recovery (Domestic, Industrial or MSW), potential demand, recovery percentage and annual economic values associated with reuse options based on baseline prices (Table S4 in OSI).

Parameter	Recoverable resources per sector and potential demand							Total revenues (million US $/year)
	Domestic WWT	Industrial WWT	MSW	Total recovery	Potential demand	Unit	Recovery percentage	
Compost	91	-	351	442	1240	ktonne/y	36%	44.2
Plastic	-	-	228	228	366	ktonne/y	62%	45.5
Paper	-	-	193	193	1185	ktonne/y	16%	38.6
Electricity	27	-	89	116	78.8	GWh/y	147%	11.6
Water	-	43	-	43	563	Mm^3/y	8%	2.6
Struvite	4.2	-	-	4.2	35	ktonne/y	12%	4.1
Total economic value								146.6

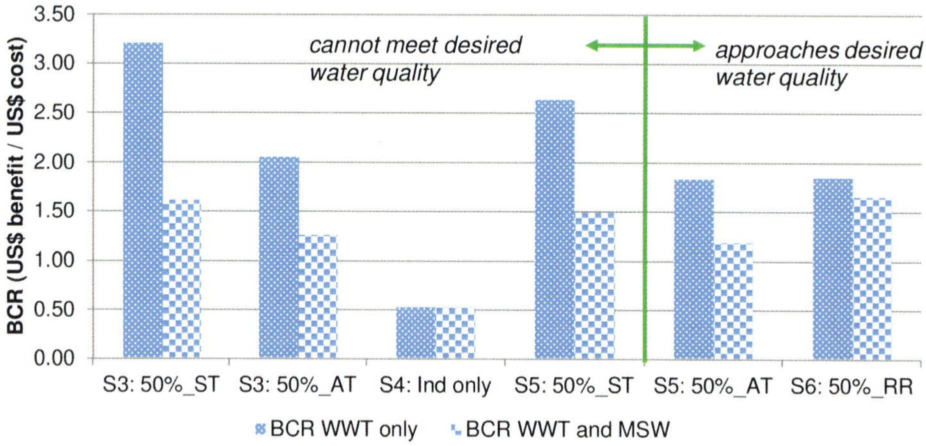

Figure 7. Calculated BCR per analyzed scenarios, differentiating the BCR in which only wastewater treatment (WWT) interventions are considered and the BCR that considers both WWT and municipal solid waste (MSW) interventions. Scenarios that approach the target water quality are S5: 50%-AT and S6: 50%-RR.

tackling both domestic and industrial pollution results in a BCR ranging from 1.83 (S5: 50% AT) to 2.64 (S5: 50% ST). However, simple technologies were not found sufficient to improve the water quality to levels approaching class II, especially in terms of nutrient (N, P) removal (**Figure 5**).

The economic returns on combined wastewater and solid waste interventions are lower than the returns on wastewater interventions only (**Figure 7**). Economic costs related to absence of solid waste services are associated with unhygienic living conditions [31], loss of tourism developments or value of land [61]. Loss of land value, however, contributes to only a fraction (4%) of total related economic impact (**Figure 6**) and on their own do not outweigh the estimated costs (see Table S6 in OSI) to establish the MSW management systems. The willingness of households to pay for solid waste collection and treatment services has been better established compared to wastewater services in Indonesia [62]. This may be attributed to direct visibility of improving solid waste management [63]. Consequently, there is a larger potential for recovering some of the costs through MSW tariffs paid by households compared with tariffs for wastewater services. Potential revenues from fees were excluded from the BCR analysis, but are relevant for development of a cost-effective wastewater and solid waste management system.

Additional benefits of resource recovery from MSW can be a driver for improving water quality. The BCR (including MSW) of scenario 5 (applying AT) is 1.19 and will increase to 1.65 by applying resource recovery (**Table 8**). The BCR of scenario 6 with MSW recovery is even higher than the BCR of Scenario 5 applying ST (1.49) showing that required additional costs to improve the water quality can be financed through the sale of resources recovered from solid waste. However, application of resource recovery from wastewater only results in a minor increase in BCR (from 1.83 to 1.85) compared to applying only advanced technology. Thus, from a financial perspective using existing market prices, the additional investments to recover resources from wastewater out

Table 8. Calculated Benefits Costs Ratio (BCR) and five alternative BCR's distinguishing (A) only WWT based BCR or (B) WWT and MSW based BCR.

		A. WWT costs and benefits						B. WWT and MSW costs and benefits					
Category	Sub category	S3: 50%_ST	S3: 50%_AT	S4: industrial only	S5: 50%_ST	S5: 50%_AT	S6: 50%_RR	S3: 50%_ST	S3: 50%_AT	S4: Industrial only	S5: 50%_ST	S5: 50%_AT	S6: 50%_RR
BCR	Baseline BCR	3.20	2.06	0.52	2.64	1.83	1.85	1.62	1.26	0.52	1.49	1.19	1.65
	Resource prices 50% of baseline	3.20	2.06	0.52	2.64	1.83	1.79	1.62	1.26	0.52	1.49	1.19	1.37
	Health impact 50% of baseline	2.22	1.42	0.52	1.86	1.29	1.34	1.12	0.87	0.52	1.05	0.84	1.33
	40 year capital lifespan	4.94	3.01	0.60	3.80	2.58	2.61	2.11	1.65	0.60	1.91	1.54	2.18
	15 year capital lifespan	2.60	1.70	0.48	2.19	1.53	1.55	1.40	1.09	0.48	1.29	1.03	1.42
	Access time gained 50% of baseline	2.75	1.76	0.52	2.28	1.58	1.61	1.39	1.08	0.52	1.28	1.03	1.50

weigh the benefits by a small margin.

In case recovered resources are sold at only half the current market price (**Table 8**) the BCR of resource recovery (S6) is lower than for AT, but still higher than 1. The BCR may change depending on the lifespan of capital stock (**Table 8**). A lifespan of 40 years results in a BCR approaching 5 (S3: 50%_ST). A major part of the cost (Table S1 in OSI) is related to sewer system developments that have typically much longer potential lifespans (even up to 100 years) [64] and therefore it is likely the BCR will be higher than the baseline BCR of 3.2 for that same scenario (S3: 50%_ST).

4. Discussion

4.1. Added Value of Integrated Approaches

Evaluating the economic performance of wastewater and solid waste interventions is a complex process, involving many variables and alternative combinations and coverage levels of interventions. Therefore a methodology was developed that combines several assessment methods and data sources in order to support decision making. The added value of the integrated approach allows for a nuanced view on interrelations compared to single cause-effect relations [65]. Thus the effects of different interventions on water quality, resource recovery potential, and related economic returns could be evaluated in parallel (**Figures 5-7**). This parallel evaluation provides significant benefits in a dynamic context [66]. It also addresses the need for a method that can quantitatively evaluate a set of sanitation alternatives to resolve trade-offs across sustainability dimensions (social, environmental, and economic) [67].

4.2. Added Value of the Approach in Practical Applications

The practical application of the integrated approach is first demonstrated in the analysis of contribution of pollution per sector (industry, domestic or agriculture) related to the pollution prevention costs. The large contribution of domestic pollution was unambiguous and confirmed in a sensitivity analysis (see also OSI, Section 7). Presented results are in line with findings of [13] who determined that households contributed 55%, industries 40% and agriculture 6% of BOD pollution entering the Saguling reservoir. Reference [11] demonstrated the importance of fertilizer use management to avoid future coastal eutrophication in Indonesian Rivers, which corresponds with the large nutrient load as a result of agricultural activities (25% for P) determined in the current study. Despite a relative low (28%) contribution of industrial COD pollution, 35% of COD can be reduced by industrial interventions, whereas the investment costs for industrial interventions are less than 10% of the domestic interventions (Table S6 in OSI). Further, the number of industries is only a fraction (~1%) of the number of households in the Citarum area and monitoring interventions would be much more practical than monitoring individual household connections. Thus, although COD pollution from industry is relatively small, it is more cost effective (>factor 5) than domestic, which may help a policy maker in prioritizing interventions.

Secondly, the integrated approach supports determination of cost-effective interventions. The added value of applying more advanced technologies or switching more people to a sewer system showed that required additional investments can be justified from the point of nutrient removal, but less so from COD removal (**Figure 4**). In addition, the use of software tools like RIBASIM to model and estimate the impact of discharged pollution loads on the anticipated water quality allows the policy maker to relate interventions and their cost to applicable water quality standards.

Thirdly, linking the resource recovery potential and its revenues to its potential demand may benefit formulation of policies or increase government involvement to foster financial sustainability of sanitation facilities [24]. The value of recoverable resources from solid waste has resulted in a very active, but informal waste recovery sector in Indonesia [68] [69]. In addition, the demonstrated potential recovery of resources exceeding the agricultural demand allows for selective marketing, focusing on safe reuse (e.g. on non-edible crops) [70] [71]. Electricity production from the joint wastewater and solid waste facilities is potentially higher than the demand for domestic wastewater and supports the potential for a joint development of wastewater and solid waste facilities [72].

Fourthly, monetizing both direct use and indirect non-use values of sanitation implementation in relation to achievable surface water quality enables the formulation of a cost and environmental effective approach. The performed analysis demonstrated that the most cost effective scenario (S3: 50%_ST) with the highest BCR differs from the scenario reaching the required water quality (e.g. S5: 50%_AT). Therefore, a policy maker needs to prioritize between these two options. As a cost effective strategy, application of advanced technologies may be restricted to the most highly densely populated urban areas (where most pollution is produced). Alternatively, a phased approach in which first simple (low cost) technologies are implemented that are later replaced, converted or extended by systems that allow for nutrient removal [73]. Monetizing benefits may further help to raise funds from other sources or actors that benefit from improved water quality, such as residential project developers or tourism sites [32].

The outcomes of the study were formulated in a planning document for the Indonesian government [53] and confirmed our hypothesis that quantification of tangible outputs using the presented approach can support policy-makers in the field.

4.3. Options for Extending the Approach

The presented framework can be further extended given the following considerations:

- To assess the sustainability of interventions and ensure that pollution is being removed and not displaced, environmental emissions other than water pollution (COD, N, P), such as odor or greenhouse gasses may be included. The effect of greenhouse gasses emitted by low cost technologies (e.g. anaerobic filters or septic tanks) is excluded from the current evaluation.
- In the determination of the water quality, several assumptions were made that may affect obtained results and could be incorporated in a next phase (see also OSI Section 9). First, a connection between surface and ground water was assumed in which infiltrated septic tank effluent load directly influences the surface water quality. Second, RIBASIM model disregards biological conversion of pollutants in the surface water, whereas these are observed in the field [12]. Third, all interventions are assumed to be designed, constructed, operated and maintained correctly, which may be optimistic in view of current practice [37] [74]. Fourth, the effect of dumped solid waste on water quality is excluded. Finally, surface water pollution from animal manure was excluded.
- The low BCR of industrial interventions (0.52, **Table 8**) and the weak mandatory industrial regulation in Indonesia [75] may suggest limited possibilities to implement industrial pollution prevention. However, alternative means to spur Indonesian industries to comply with environmental standards such as public disclosure (the regular collection and dissemination of information about firms' environmental performance) have been shown to be effective [76].
- Aerobic technologies were used as industrial references, whereas the use of anaerobic technologies may result in lower investment and/or operational costs [77] [78].
- Not all economic impacts were quantified in this study (see also OSI, Section 10), such as consumption of fish imbibing toxic wastes or otherwise infected [79], reduced land subsidence and improved recreational values [30] [80]. In addition, long-term impacts on the river and population of industrially discharged toxins and heavy metals were excluded and would specifically increase the BCR of scenario S4 (industrial intervention). Although the current study focuses on a part of a river basin only and uses locally collected data, a further detailing on a city or community level may be required to prevent overgeneralization and misunderstanding of individuals' preference tradeoffs [81].

- Applying advanced technologies (AT) will improve water quality (**Figure 5**), but will not increase quantified health or welfare impact. At the same, anticipated long-term effects of reduced eutrophication and less impacted ecosystem functioning [11] were not quantified, whereas these would further increase the BCR.
- The BCR considers the overall societal perspective, whereas different costs and benefits are incurred and enjoyed by different stakeholders. Thus, the costs of domestic interventions are to a large extent paid for by the national and local governments (in Indonesia ~ 70%) and to lesser extent by individual households [50], whereas industries typically pay the costs of the interventions themselves [37]. Benefits of improved water quality as a result of interventions can be either increased revenues (e.g. sale of recovered resources) or averted costs (e.g. lower water treatment costs) which benefit a single party, or are generalized to the population (e.g. averted health or time costs) which benefit society as a whole [30]. In the elaboration of a planning document, the incidence of costs and benefits should be further detailed. In addition, serious institutional challenges have been identified in providing access to sanitation in a development world's context [82]. To implement the planned sanitation development, the responsible institutions need to be identified and budgets should be allocated. A methodology to do this, involving private households, local and national governments, was described by [50].

5. Conclusions

In this study, an integrated method was presented that quantified the economic costs and benefits of wastewater and solid waste interventions in relation to water quality improvements and resource recovery potential. The approach provides added value in the decision making process in a complex and dynamic context since it helps resolve trade-offs across different dimensions of sustainability (e.g. social, environmental and economic).

Identification of pollution sources and the impact of interventions on discharged pollution loads allows for prioritizing of actions. By simultaneously modeling the water quality and cost impact of variations in 1) type of technology and 2) the household numbers switching from poor-performing septic tanks to off-site systems, insight into the cost-effectiveness of environmental policies is provided. This allows a policy maker to optimize economic and water quality benefits.

In the presented case of the Upper Citarum River, domestic interventions applying simple technologies were most attractive, with an estimated BCR of 3.2. However, to achieve the target water quality both industrial and advanced domestic WWT technologies would be required, leading to an estimated BCR of 2.0. Resource recovery from MSW was found to be a driver for improving water quality, as benefits through the sale of recovered resource outweighed the additional costs to improve the water quality.

Acknowledgements

The authors acknowledge the contribution of Aart van Nes, Fery Hardianto and Wil van der Krogt in the data collection and RIBASIM modeling as part of the 6 Ci's project funded by the ADB (TA7189-INO: Institutional Strengthening for Integrated Water Resources Management (IWRM) in the 6 Ci's River Basin Territory-Package B). The economic works of this study was funded by WSP's Multi-Donor Trust Fund for WSP East Asia and the Pacific, supported by the Government of Australia. Guy Hutton conducted the work while employed at the World Bank. We thank Isabel Blackett, Almud Weitz, Enrico Rahadi Djonoputro, and Deviariandy Setiawan for their valuable input. In loving memory of pak Nugroho-Director of Urban, Housings and Settlements, National Development Planning Agency (Bappenas); may you rest in peace.

References

[1] WHO and UNICEF (2015) Progress on Sanitation and Drinking Water—2015 Update and MDG Assessment. Geneva, Switzerland.

[2] Corcoran, E., Nellemann, C., Baker, E., Bos, R., Osborn, D. and Savelli, H. (2010) Sick Water? The Central Role of Wastewater Management in Sustainable Development. A Rapid Response Assessment. United Nations Environment Programme, UN-HABITAT, GRID-Arendal.

[3] ADB (2013) Indonesia: Water Supply and Sanitation Sector Assessment, Strategy, and Road Map. Manilla, Philippines.

[4] ADB (2014) Urban Water Supply and Sanitation in Southeast Asia: A Guide to Good Practice. Manilla, Philippines.

[5] Shuval, H. (2003) Estimating the Global Burden of Thalassogenic Diseases: Human Infectious Diseases Caused by Wastewater Pollution of the Marine Environment. *Journal of Water and Health*, **1**, 53-64.

[6] Montgomery, M.A.A. and Elimelech, M. (2007) Water and Sanitation in Developing Countries: Including Health in the Equation. *Environmental Science & Technology*, **41**, 17-24. http://dx.doi.org/10.1021/es072435t

[7] Waddington, H. and Snilstveit, B. (2009) Effectiveness and Sustainability of Water, Sanitation, and Hygiene Interventions in Combating Diarrhoea. *Journal of Development Effectiveness*, **1**, 295-335. http://dx.doi.org/10.1080/19439340903141175

[8] Surinkul, N. and Koottatep, T. (2009) Advanced Sanitation Planning Tool with Health Risk Assessment: Case Study of a Peri-Urban Community in Thailand. *Human and Ecological Risk Assessment*, **15**, 1064-1077. http://dx.doi.org/10.1080/10807030903153469

[9] WSP (2013) Assessment of Sludge Accumulation and Pit Filling Rates in Indonesia. Jakarta, Indonesia.

[10] Wright, J., Cronin, A., Okotto-Okotto, J., Yang, H., Pedley, S. and Gundry, S.W. (2013) A Spatial Analysis of Pit Latrine Density and Groundwater Source Contamination. *Environmental Monitoring and Assessment*, **185**, 4261-4272. http://dx.doi.org/10.1007/s10661-012-2866-8

[11] Suwarno, D., Löhr, A., Kroeze, C. and Widianarko, B. (2013) Past and Future Trends in Nutrient Export by 19 Rivers to the Coastal Waters of Indonesia. *Journal of Integrative Environmental Sciences*, **10**, 55-71. http://dx.doi.org/10.1080/1943815X.2013.772902

[12] Hart, B.T., van Dok, W. and Djuangsih, N. (2002) Nutrient Budget for Saguling Reservoir, West Java, Indonesia. *Water Research*, **36**, 2152-2160. http://dx.doi.org/10.1016/S0043-1354(01)00428-6

[13] Suharyanto and Matsushita, J. (2011) Integrated Basin-Based Wastewater Management System for Water Pollution Control in an Enclosed Waterbody of the Upper Citarum River Basin, Indonesia: Case study of Saguling Reservoir. *Lakes & Reservoirs: Research & Management*, **16**, 185-193. http://dx.doi.org/10.1111/j.1440-1770.2011.00473.x

[14] Fulazzaky, M.A. (2010) Water Quality Evaluation System to Assess the Status and the Suitability of the Citarum River Water to Different Uses. *Environmental Monitoring and Assessment*, **168**, 669-684. http://dx.doi.org/10.1007/s10661-009-1142-z

[15] Ujang, Z. and Henze, M. (2006) Municipal Wastewater Management in Developing Countries: Principles and Engineering. IWA Publishing, Illustrate.

[16] MoPW (2011) TA7189-INO: Institutional Strengthening for Integrated Water Resources Management (IWRM) in the 6 Ci's River Basin Territory-Package B. Jakarta, Indonesia.

[17] Abu Ghunmi, L., Zeeman, G., Fayyad, M. and van Lier, J.B. (2011) Grey Water Treatment Systems: A Review. *Critical Reviews in Environmental Science and Technology*, **41**, 657-698. http://dx.doi.org/10.1080/10643380903048443

[18] Hatt, B.E., Fletcher, T.D., Walsh, C.J. and Taylor, S.L. (2004) The Influence of Urban Density and Drainage Infrastructure on the Concentrations and Loads of Pollutants in Small Streams. *Environmental Management*, **34**, 112-124. http://dx.doi.org/10.1007/s00267-004-0221-8

[19] Kerstens, S.M., Leusbrock, I. and Zeeman, G. (2015) Feasibility Analysis of Wastewater and Solid Waste Systems for Application in Indonesia. *Science of the Total Environment*, **530-531**, 53-65. http://dx.doi.org/10.1016/j.scitotenv.2015.05.077

[20] Antonopoulos, I.-S., Perkoulidis, G., Logothetis, D. and Karkanias, C. (2014) Ranking Municipal Solid Waste Treatment Alternatives Considering Sustainability Criteria Using the Analytical Hierarchical Process Tool. *Resources, Conservation and Recycling*, **86**, 149-159. http://dx.doi.org/10.1016/j.resconrec.2014.03.002

[21] Orhon, D., Germirli, F. and Karahan, O. (2009) Management of Industrial Wastewaters. Industrial Wastewater Treatment by Activated Sludge, First. IWA Publishing, London, 400.

[22] Gozálvez-Zafrilla, J.M., Sanz-Escribano, D., Lora-García, J. and León Hidalgo, M.C. (2008) Nanofiltration of Secondary Effluent for Wastewater Reuse in the Textile Industry. *Desalination*, **222**, 272-279. http://dx.doi.org/10.1016/j.desal.2007.01.173

[23] Abery, N.W., Sukadi, F., Budhiman, A.A., Kartamihardja, E.S. and Koeshendrajana, S. (2005) Fisheries and Cage Culture of Three Reservoirs in West Java, Indonesia, a Case Study of Ambitious Development and Resulting Interactions. *Fisheries Management and Ecology*, **12**, 315-330. http://dx.doi.org/10.1111/j.1365-2400.2005.00455.x

[24] Murray, A. and Ray, I. (2010) Commentary: Back-End Users: The Unrecognized Stakeholders in Demand-Driven Sanitation. *Journal of Planning Education and Research*, **30**, 94-102. http://dx.doi.org/10.1177/0739456X10369800

[25] Kerstens, S.M., De Mes, T.Z.D. and Lue, B. (2009) Designing Sustainable Sanitation in Urban Planning Proposed for Changzhou, China. *Water, Science and Technology*, **60**, 3165-3172. http://dx.doi.org/10.2166/wst.2009.160

[26] Aye, L. and Widjaya, E.R. (2006) Environmental and economic analyses of waste disposal options for traditional markets in Indonesia. *Waste Management*, **26**, 1180-1191. http://dx.doi.org/10.1016/j.wasman.2005.09.010

[27] Aprilia, A., Tezuka, T. and Spaargaren, G. (2012) Household Solid Waste Management in Jakarta, Indonesia: A Socio-Economic Evaluation. In: Rebellon LFM, Ed., *Waste Management—An Integrated Vision*. InTech, Croatia, 30. http://dx.doi.org/10.5772/51464

[28] Nandy, B., Sharma, G., Garg, S., Kumari, S., George, T., Sunanda, Y., *et al.* (2015) Recovery

of Consumer Waste in India—A Mass Flow Analysis for Paper, Plastic and Glass and the Contribution of Households and the Informal Sector. *Resources, Conservation and Recycling*, **101**, 167-181. http://dx.doi.org/10.1016/j.resconrec.2015.05.012

[29] Kerstens, S.M., Priyanka, A., Van Dijk, K.C., De Ruijter, F.J., Leusbrock, I. and Zeeman, G. (2016) Potential Demand for Recoverable Resources from Indonesian Wastewater and Solid Waste. *Resources, Conservation and Recycling*, **110**, 16-29. http://dx.doi.org/10.1016/j.resconrec.2016.03.002

[30] Alam, K. (2008) Cost-Benefit Analysis of Restoring Buriganga River, Bangladesh. *International Journal of Water Resources Development*, **24**, 593-607. http://dx.doi.org/10.1080/07900620802224486

[31] Winara, A., Hutton, G., Oktarinda Purnomo, E., Hadiwardoyo, K., Merdykasari, I., *et al.* (2011) Economic Assessment of Sanitation Interventions in Indonesia. Jakarta, Indonesia.

[32] Hutton, G. (2013) Global Costs and Benefits of Reaching Universal Coverage of Sanitation and Drinking-Water Supply. *Journal of Water and Health*, **11**, 1-12. http://dx.doi.org/10.2166/wh.2012.105

[33] Hutton, G., Rodriguez, U.-P., Winara, A., Anh, N.V., Phyrum, K., Chuan, L., *et al.* (2014) Economic Efficiency of Sanitation Interventions in Southeast Asia. *Journal of Water, Sanitation and Hygiene for Development*, **4**, 23-26. http://dx.doi.org/10.2166/washdev.2013.158

[34] Ward, F.A. (2012) Cost-Benefit and Water Resources Policy: A Survey. *Water Policy*, **14**, 250-280. http://dx.doi.org/10.2166/wp.2011.021

[35] Ng, B.J.H., Zhou, J., Giannis, A., Chang, V.W.-C. and Wang, J.-Y. (2014) Environmental Life Cycle Assessment of Different Domestic Wastewater Streams: Policy Effectiveness in a Tropical Urban Environment. *Journal of Environmental Management*, **140**, 60-68. http://dx.doi.org/10.1016/j.jenvman.2014.01.052

[36] Malekpour, S., Langeveld, J., Letema, S., Clemens, F. and van Lier, J.B. (2013) Judgment under Uncertainty: A Probabilistic Evaluation Framework for Decision-Making about Sanitation Systems in Low-Income Countries. *Journal of Environmental Management*, **118**, 106-114. http://dx.doi.org/10.1016/j.jenvman.2013.01.007

[37] De Vries, D. (2012) BSc Thesis: Feasible Wastewater Treatment Technologies for Upper Citarum's Textile Companies. Wageningen University, Wageningen.

[38] Deltares (2009) RIBASIM River Basin Planning and Management Software Guide. Deltares, Delft, 4.

[39] BPLHD (2010) Water Quality Data for COD, BOD, N and P for the Period 2001-2009 in the Upper Citarum River at Different Locations from West Java Regional Control Agency of the Living Environment (BPLHD). Received as XLS Files as Part of the ADB's 6 Ci's Project TA7189.

[40] BPLHD West Java (2011) River Water Quality in Bandung in 2011 [Original Title in Indonesia: kualitas air sungai prioritas prokasih 2011]. Bandung.

[41] DKI (2005) Wastewater Management in Province DKI [Originally Title in Indonesian: Pengelolaan air limbah domestik di provinsi daerah khusus ibukota jakarta dengan]. Jakarta, Indonesia, 23.

[42] Bojong Soang (2012) Bojong Soang WWTP Access Data Obtained From a Visit to the Plant.

[43] ESDM (2009) Industrial Ground Water Abstraction Values (Base Year 2009) in the Upper Citarum River Area Obtained from the Energy and Mineral Resources Department (ESDM).

[44] PSDA (2010) Industrial Surface Water Abstraction Data (2010) Obtained Directly from the Provincial Agency for Water Resources Management (PSDA).

[45] BPLHD (2010) Report of the Results of Self Monitoring Wastewater and Air Quality Activities in West Java. Environmental Pollution Monitoring Activities Fiscal Year 2010 [Original Title in Indonesia: Laporan hasil self monitoring limbah cair & kualitas udara pada kegiat]. Bandung.

[46] KNLH (2010) Industry Monitoring Report Recommendations. Chapter II. Monitoring of Industrial Activity [Original Title in Indonesia: Laporan Rekomendasi Pemantauan Industri, BAB II. Pemantauan Kegiatan Industri]. Cimahi.

[47] BWRP (2000) Guidelines for Basin Water Resources Planning—Water Quality. Basin Water Resources Planning Project.

[48] Arantes, V. and Milagres, A.M.F. (2007) Application of Statistical Experimental Design to the Treatment of Bleaching Kraft Mill Effluent Using a Mediated Free Radical System. *Water, Science and Technology*, **55**, 1-7. http://dx.doi.org/10.2166/wst.2007.205

[49] Pronk, M., de Kreuk, M.K., de Bruin, B., Kamminga, P., Kleerebezem, R. and van Loosdrecht, M.C.M. (2015) Full Scale Performance of the Aerobic Granular Sludge Process for Sewage Treatment. *Water Research*, **84**, 207-217. http://dx.doi.org/10.1016/j.watres.2015.07.011

[50] Kerstens, S.M., Spiller, M., Leusbrock, I. and Zeeman, G. (2016) A New Approach to Nationwide Sanitation Planning for Developing Countries: Case Study of Indonesia. *Science of the Total Environment*, **550**, 676-689. http://dx.doi.org/10.1016/j.scitotenv.2016.01.104

[51] MPS and Nijhuis (2012) Discussions with International Suppliers in Indonesia (MPS and Nijhuis) on Applicable WWTP for Textile Wastewater in Indonesia.

[52] Gonenc, I.E., Wolflin, J.P. and Russo, R.C. (2014) Sustainable Watershed Management. CRC Press, London. http://dx.doi.org/10.1201/b17433

[53] Kerstens, S.M., Hutton, G., Van Nes, A. and Firmansyah, I. (2013) Downstream Impacts of Water Pollution in the Upper Citarum River, West Java, Indonesia Economic Assessment of Interventions to Improve Water Quality. Jakarta, Indonesia.

[54] Waddington, H., Snilstveit, B., White, H. and Fewtrell, L. (2009) Water, Sanitation and Hygiene Interventions to Combat Childhood Diarrhoea in Developing Countries. The International Initiative for Impact Evaluation (3IE).

[55] Barreto, M.L., Genser, B., Strina, A., Teixeira, M.G., Assis, A.M.O., Rego, R.F., *et al.* (2010) Impact of a Citywide Sanitation Program in Northeast Brazil on Intestinal Parasites Infection in Young Children. *Environmental Health Perspectives*, **118**, 1637-1642. http://dx.doi.org/10.1289/ehp.1002058

[56] Moraes, L.R.S., Cancio, J.A., Cairncross, S. and Huttly, S. (2003) Impact of Drainage and Sewerage on Diarrhoea in Poor Urban Areas in Salvador, Brazil. *Transactions of the Royal Society of Tropical Medicine and Hygiene*, **97**, 153-158. http://dx.doi.org/10.1016/S0035-9203(03)90104-0

[57] Gwilliam, K.M. (1997) The Value of Time in Economic Evaluation of Transport Projects: Lessons from Recent Research. Washington DC, Report No. OT-5.

[58] WSP (2012) Interview with Field Officers of the Provincial Fishery Department of West Java Performed by WSP. Bandung.

[59] Ministry of Health (2010) Riset Kesehatan Dasar [Basic Health Research]. Jakarta, Indonesia.

[60] GoI (2001) Presidential Decree No 82, 2001 Water Quality Management and Water Pollution Control [Pengelolaan kualitas air dan pengendalian pencemaran air]. Government of Indonesia, Jakarta, Indonesia, 1-31.

[61] Alam, M.K. and Marinova, D. (2003) Measuring the Total Value of a River Cleanup. *Water,*

Science and Technology, **48**, 149-156.

[62] WSP (2011) Lessons in Urban Sanitation Indonesia Sanitation Sector Development: Indonesia Sanitation Sector Development Program 2006-2010. Jakarta, Indonesia.

[63] Winters, M.S., Karim, A.G. and Martawardaya, B. (2014) Public Service Provision under Conditions of Insufficient Citizen Demand: Insights from the Urban Sanitation Sector in Indonesia. *World Development*, **60**, 31-42. http://dx.doi.org/10.1016/j.worlddev.2014.03.017

[64] Petit-Boix, A., Sanjuan-Delmás, D., Gasol, C.M., Villalba, G., Suárez-Ojeda, M.E., Gabarrell, X., *et al.* (2014) Environmental Assessment of Sewer Construction in Small to Medium Sized Cities Using Life Cycle Assessment. *Water Resources Management*, **28**, 979-997. http://dx.doi.org/10.1007/s11269-014-0528-z

[65] Mirakyan, A. and De Guio, R. (2013) Integrated Energy Planning in Cities and Territories: A Review of Methods and Tools. *Renewable and Sustainable Energy Reviews*, **22**, 289-297. http://dx.doi.org/10.1016/j.rser.2013.01.033

[66] Pollack, J. (2009) Multimethodology in Series and Parallel: Strategic Planning Using Hard and Soft OR. *Journal of the Operational Research Society*, **60**, 156-167. http://dx.doi.org/10.1057/palgrave.jors.2602538

[67] Guest, J.S., Skerlos, S.J., Barnard, J.L., Beck, M.B., Daigger, G.T., Hilger, H., *et al.* (2009) A New Planning and Design Paradigm to Achieve Sustainable Resource Recovery from Wastewater. *Environmental Science and Technology*, **43**, 6126-6130. http://dx.doi.org/10.1021/es9010515

[68] Chaerul, M., Fahruroji, A.R. and Fujiwara, T. (2013) Recycling of Plastic Packaging Waste in Bandung City, Indonesia. *Journal of Material Cycles and Waste Management*, **16**, 509-518. http://dx.doi.org/10.1007/s10163-013-0201-2

[69] Sasaki, S. and Araki, T. (2013) Employer-Employee and Buyer-Seller Relationships among Waste Pickers at Final Disposal Site in Informal Recycling: The Case of Bantar Gebang in Indonesia. *Habitat International*, **40**, 51-57. http://dx.doi.org/10.1016/j.habitatint.2013.02.003

[70] WHO (2006) Guidelines for the Safe Use of Wastewater, Excreta and Greywater. Volume 1 Policy and Regulatory Aspects. WHO (World Health Organization) Press, Geneva, Switzerland.

[71] Reijnders, L. (2014) Phosphorus Resources, Their Depletion and Conservation: A Review. *Resources, Conservation and Recycling*, **93**, 32-49. http://dx.doi.org/10.1016/j.resconrec.2014.09.006

[72] Zitomer, D.H., Adhikari, P., Heisel, C. and Dineen, D. (2008) Municipal Anaerobic Digesters for Codigestion, Energy Recovery, and Greenhouse Gas Reductions. *Water Environment Research*, **80**, 229-237. http://dx.doi.org/10.2175/106143007X221201

[73] Spiller, M., Vreeburg, J.H.G., Leusbrock, I. and Zeeman, G. (2015) Flexible Design in Water and Wastewater Engineering—Definitions, Literature and Decision Guide. *Journal of Environmental Management*, **149**, 271-281. http://dx.doi.org/10.1016/j.jenvman.2014.09.031

[74] ADB (2013) TA-7739 INO: Supporting Water Operators' Partnerships. Indonesia Water Supply and Sanitation Institute (IWWI) (44090-012), Jakarta, Indonesia.

[75] D'Hondt, L. (2013) Indonesia's Environmental Law of 2009 and Its Administrative Coercion Provisions. A Conceptual Misunderstanding with Large Practical Implications? SSRN (Social Science Research Network), Leiden, 1-7.

[76] Blackman, A. (2010) Alternative Pollution Control Policies in Developing Countries. *Review of Environmental Economics and Policy*, **4**, 234-253.

http://dx.doi.org/10.1093/reep/req005

[77] Rajeshwari, K., Balakrishnan, M., Kansal, A., Lata, K. and Kishore, V.V. (2000) State-of-the-Art of Anaerobic Digestion Technology for Industrial Wastewater Treatment. *Renewable and Sustainable Energy Reviews*, **4**, 135-156. http://dx.doi.org/10.1016/S1364-0321(99)00014-3

[78] Van Lier, J.B. (2008) High-Rate Anaerobic Wastewater Treatment: Diversifying from End-of-the-Pipe Treatment to Resource-Oriented Conversion Techniques. *Water, Science and Technology*, **57**, 1137-1148. http://dx.doi.org/10.2166/wst.2008.040

[79] Lasut, M.T., Jensen, K.R. and Shivakoti, G. (2008) Analysis of Constraints and Potentials for Wastewater Management in the Coastal City of Manado, North Sulawesi, Indonesia. *Journal of Environmental Management*, **88**, 1141-1150. http://dx.doi.org/10.1016/j.jenvman.2007.06.011

[80] Day, B. and Mourato, S. (1998) Willingness to Pay for Water Quality Maintenance in Chinese Rivers. CSERGE Working Paper WM 98-02, London.

[81] Whittington, D., Jeuland, M., Barker, K. and Yuen, Y. (2012) Setting Priorities, Targeting Subsidies among Water, Sanitation, and Preventive Health Interventions in Developing Countries. *World Development*, **40**, 1546-1468. http://dx.doi.org/10.1016/j.worlddev.2012.03.004

[82] McGranahan, G. (2015) Realizing the Right to Sanitation in Deprived Urban Communities: Meeting the Challenges of Collective Action, Coproduction, Affordability, and Housing Tenure. *World Development*, **68**, 242-253. http://dx.doi.org/10.1016/j.worlddev.2014.12.008

14

The Contribution of Environmental Impact Assessment Studies to the Flora Biodiversity in Sudan

Osman M. M. Ali, Mohamed F. Idris

Institute of Environmental Studies, Khartoum, Sudan
Email: osmirghani@yahoo.co.uk

Abstract

Environmental Impact Assessment Studies (EIA) for four developmental projects were chosen for this study viz-a-viz., oil and gas, river diversion, hydropower and gold mining. The EIA studies for these activities have been conducted in three different ecological regions: the flood region in the former southern Sudan, low rainfall Savannah and desert region. Trees, shrubs and herbaceous species were identified and recorded in each region. Species encountered for the first time, endangered, invasive and increasers were also reported. The EIA studies have produced lists of natural flora in areas which were difficult to access without the EIA missions. During the EIA study for Jonglei Canal project the new mono specific genus *Suddia sagitifolia* was discovered in the flood region as the only endemic aquatic plant in Sudan. Within the same region, the smallest floating plant in Sudan (*Wollfia hyaline*) was reported. The critically endangered "Argon Palm" (*Medemia argun*) was encountered in the Nubian Desert of the Red Sea State during a gold mining EIA study. The Seha (*Belpharis persica*), not reported to exist in desert conditions, has been found in the Great Desert west of the Nile during an EIA study for Kajbar Hydropower project. The paper calls for the publication of the above findings and highlights the need for further detailed studies in order to enhance knowledge of the Sudanese biodiversity status.

Keywords

Environmental Impact Assessment, Flora, Biodiversity, Endangered

1. Introduction

Sudan is endowed with many ecological zones; they extend from the desert and semi-desert in the north to the low rainfall woodland savannah in the south. The ecological variability made Sudan rich in plants biodiversity encompassing many agricultural crops, fruit trees and natural vegetation of woody and herbaceous plants.

The flora diversity in the natural ecosystems supplies the country population-nomadic pastoralists, agro-pastoralists and urban communities with a wide range of services and values such as:

- Source of food and feed for humans and animals;
- Maintenance of the gaseous quality of the atmosphere;
- Mitigation of global warming via contribution in carbon sequestration;
- Protection of watershed areas and buffering against floods impacts;
- Provision of medicinal and aromatic products;
- Contribution in soil formation;
- Protection and maintenance of the productive capacity of the lands by reducing water and wind force actions; and
- Participating in cultural and aesthetic values.

However, the natural plants are under continuous threats emanating from, *inter alia*:

- Overgrazing;
- Cutting for cropping, fuel and building;
- Desertification and climate change; and
- Land preparation for development projects.

The situation is aggravated by lack of a complete, authentic and updated list of the natural flora. Lack of experts, funds and the presence of some of these plant communities in remote, inaccessible regions have exacerbated the grave situation. It is for the latter constraint that environmental impact assessment studies have come to play a significant alleviating role.

EIA is a systematic process to identify, predict and evaluate the environmental effects of proposed actions and projects. The process also proposed appropriate mitigation measures to offset significant adverse impacts of the proposed projects [1]. To fulfil the mandate of an EIA and for its objectives to be materialized, the environment where the project is going to be constructed and operated, or where its impacts could reach, the original environment has to be studied. Such a study which entails the documentation of present status of the physical and biological environmental components is called the baseline study. The geology, hydrology, (surface and ground water), meteorology, air quality, soil, vegetation, wildlife, land use and socio-economic aspects are all components of the baseline part of the EIA.

2. Material and Methods

Four developmental projects were chosen for this study viz-a-viz., oil and gas, hydropower, river diversion and gold mining. The EIA studies for these activities have been conducted in three ecological regions: the flood region in the former southern Sudan, low rainfall Savannah and in the desert region, respectively.

Qualitative and quantitative analyses of the vegetation have been carried out in each region selected sites. The plants encountered were qualitatively described into trees, shrubs and herbaceous species. The flora was identified and recorded in each region. Species encountered for the first time, new and endangered species were also reported. The identification resulted in the preparation of a list of grasses, other herbs (forbs) on the herbaceous layer, as well as trees and shrubs of the woody layer in the project area. Species which were not identified in the field were photographed, samples collected and later on identified through taxonomy books and by consulting experts. Quantitative estimates were conducted for the woody species using a quadrate (20 m × 20 m) to get the density of trees and shrubs in the area. Inventory of both qualitative and quantitative parameters were done along selected transect lines covering the project area. Other observations on the sampling points and along the selected transects such as soil type and condition, topography and land use were also recorded. GPS was used to document the studied flora together with photography with a digital camera. **Tables 1-3** show the species lists of the flora in the EIA areas of studied projects.

3. Results and Discussions

3.1. Provision of Recent/Updated Plant Lists

The first list is obtained from an EIA study for an oil exploration project in Block 5B in the swamps of the upper stretch of the White Nile in the former southern Sudan WNPOC [2].

The list in **Table 1** will have to be looked upon in the light of the paucity of aquatic plants lists [3].

The second list (**Table 2**) describes the flora in the desert zone obtained during an EIA study for the proposed Kajbar Hydropower project [4]. The plants were classified into trees, shrubs, grasses and forbes.

The third list of plants (**Table 3**) was assembled during an EIA study tour for an oil development project in Western Kordofan State where a pipeline was proposed to transport crude oil [5].

The area in which the proposed oil pipeline crosses is known for its good vegetation cover especially during the rainy season. Harrison and Jackson [6] classified the vegetation of this area as *Combretum glutinosum-Dalbergia melanoxylon-Albizia amara* Savannah woodland as sub-division of the low rainfall woodland savannah on sand. This sub-division occurs in areas of rainfall ranging between 450 - 600 mm/year. The dominant woody species were mostly broad-leafed non-thorny such as: *Combretum glutinosum, D. melanoxylon, A. amara, Guirea senegalensis, Sclerocarya birrea, Commiphora africana, Lannea humilis, Terminalia laxiflora, Terminalia brownii, Dalbergia melanoxylon* and *Adansonia digitata.* Currently *C. africana* and *L. humilis* are decreasing due to their intensive use by the pastoralists and agro-pastoralists for robes making, buildings materials etc. while *G. senegalensis, C. glutinosum* and *Terminalia* spp. are increasing due to less destruction. *Dichrostachys cinerea* is found to increase along animal tracks, night camps and roads sides. *Dalbergia melanoxylon* is severely decreasing due to the high run-off, on the bare areas around the small savannah deciduous tree causing root system desiccation and also the human intensive use of the wood.

Table 1. Aquatic plants of the flood region in block B 5.

No	Botanical Name	Family
1	*Azolla nilotica*	Azollaceae
2	*Cyclosorus interruptus*	Azollaceae
3	*Cyperus papyrus*	Cyperaceae
4	*Cyperus mundii*	Cyperaceae
5	*Phragmites karka*	Poaceae
6	*Vosia cuspidate*	Poaceae
7	*Echinochloa stagnina*	Poaceae
8	*Echinochloa pyramidalis*	Poaceae
9	*Oryza barthii*	Poaceae
10	*Oryza longstminata*	Poaceae
11	*Hyparrhenia rufa*	Poaceae
12	*Seteria incrassata*	Poaceae
13	*Suddia sagitifolia**	Poaceae
14	*Vetiveria nigritana*	Poaceae
15	*Typha domingensis*	Typhaceae
16	*Luffa cylidrica*	Cucurbitaceae
17	*Ipomoea carca*	Concolvulaceae
18	*Ipomoea aquatica*	Concolvulaceae
19	*Vigna nilotica*	Papilionaceae
20	*Trapa natans*	Trapaceae
21	*Eichhornia carssipes*	Pontederiaceae
22	*Nymphaea lotus*	Nymphaceae
23	*Nymphaea caerulea*	Nymphaceae
24	*Nymphaea micrantha*	Nymphaceae
25	*Pistia stratiotes*	Araceae
26	*Lemna gibba*	Lemnaceae
27	*Walffia hyaline*	Lemnaceae
28	*Polygnoum pulchrum*	polygonaceae
29	*Ceratophyllum demersum*	Ceratophyllaceae
30	*Melanthera scadens* ssp. *madagasciensis*	Astraceae
31	*Cissus ibuensis*	Ampelidaceae
32	*Hyphaene thebaica*	Palmaceae
33	*Barassus aethiopium*	Palmaceae

*Nomen novum.

The previous ground cover was mentioned to be composed of *Brachiarra xantholeuea, Eragrostis termula, Cenchrus biflorus, Aristidapallidac, Andropogan gayanus, Pennsetium pedicellatum, Blepharis linariifolia, Zornia glochidiata* and *Monechma hispidum*. Currently, *C. biflorus* (Huskaneit) and *B. linariifolia* were not observed and the

Table 2. Natural trees and shrubs in Kajbar area, northern Sudan natural grasses and other herbs.

Botanical Name	Family
Aerva javanica	Amarnthaceae
Abutilon spp.	Malvaceae
Demostachya cynosuroides	Poaceae
Senna italica	Caesalpiniaceae
Haplophyllum tuberculatum	Rutaceae
Polygonum setulosum	Polygonaceae
Tephrosia apollinea	Fabaceae
Fagonia critica	Zygophllaceace
Indigofera ablongifolia	Fabaceae
Euphorbia aegyptiaca	Euphorbiaceae
Cynadon dactylon	Poaceae
Daicanthium annulatum	Poaceae
Rhynchosia minima	Fabaceae
Blepharis persica	Acanthaceae
Blepharisl inariifolia	Asteraceae
Solanum incanum	Solanaceae
Chrozaphora spp.	Euphorbiaceae
Xanthium barsilicum	Asteraceae
Crotalaria thebica	Fabaceae
Farsetia spp.	Cruciperae
Panicum turgidum	Poaceae
Datura stramonium	Solanaceae
Citrullus colocynthis	cucurbitaceae
Boerhavia repens	Myctaginaceae
Cyperus rotundus	Cyperaceae
Cyperus spp.	Cyperaceae

people met confirmed their disappearance. List of the current natural vegetation in the project area is given in **Table 3**.

3.2. Evidence of Sightings in Unrecorded Sites and Locations

Belpharis persica (Burm.f.) kuntze. Synonym: *Belpharis edulis*

This is an annual woody herb, with spiny leaf margin of the family Acanthaceae. The green and dry plant is a very valuable camel fodder while sheep, goats and cattle relish the plant when green. It is found in the semi-desert ecological zones of Butana plain of eastern Sudan, Erkawit and khor Dahand in the Red Sea State [7]. It was not specifically

Table 3. List of the natural vegetation in the area of the oil pipe line from Baleela to Um Adara. (a) Woody species, (b) Herbaceous species.

(a)

Scientific name	Family
Acacia nilotica	Leguminosae: Mimosoideae
Acacia gerrarrdii	Leguminosae: Mimosoideae
Acacia senegal	Leguminosae: Mimosoideae
Acacia laeta	Leguminosae: Mimosoideae
Acacia seyal	Leguminosae: Mimosoideae
Acacia mellifera	Leguminosae: Mimosoideae
Albizia amara	Leguminosae: Mimosoideae
Anogeissus leiocarpus	Combretaceae
Adansonia digitata	Bombacaceae
Balanites aegyptiaca	Zygophyllaceae
Calotropis procera	Asclepiadaceae
Dalbergia melanoxylon	Leguminosae: Papilionoideae
Dichrostachys cinerea	Leguminosae: Mimosoideae
Guiera senegalensis	Combretaceae
Maerua spp.	Capparaceae
Sclerocarya birrea	Anacardiaceae
Terminalia laxiflora	Combretaceae
Ziziphus spina-christi	Rhamnaceae
Combretum aculeatum	Combretaceae
Combretum glutinosum	Combretaceae
Combretum hartmannianum	Combretaceae
Piliostigma reticulatum	Leguminosae: Caesalpinoideae
Boswellia papyrifera	Burseraceae
Cordia Africana	Boraginaceae
Ipomoea cornea	Convolvulaceae
Ficus spp.	Moraceae
Commiphora africana	Burseraceae

(b)

Scientific name	Family
Aristida mutabilis	Poaceae
Aristida pallidae	Poaceae
Andropogon gayanus	Poaceae
Abutilon spp.	Malvaceae
Brahiaria obtusiflora	Poaceae

Continued

Eragrostis tremula	Poaceae
Cassia tora	Leguminose: Caesalpinioideae
Cassia mimosoides	Leguminose: Caesalpinioideae
Setaria pallide-fusca	Poaceae
Echinochloa colona	Poaceae
Dactyloctenium aegyptium	Poaceae
Pennisteum pedicellatum	Poaceae
Corchorus spp.	Malvaceae
Ipomeoa repens	Convolvulaceae
Solanum dubium	Solanaceae
Stylosanthes fruticosa	Leguminoseae: Papilinioideae
Indigofera spp.	Leguminoseae: Papilinioideae
Euphorbia spp.	Euphorbiaceae
Cyperus spp.	Cyperaceae
Brachiaria xantholeuca	Poaceae
Euphorbia aegyptiaca	Euphorbiaceae
Chloris spp.	Poaceae
Zornia glochidiata	Leguminoseae: Papilinioideae
Ctenium elegans	Poaceae
Digitaria intecta	Poaceae
Hyparrhenia confins	Poaceae
Sesabania sesaban	Leguminoseae: Papilinioideae
Oldenlandia senegalensis	Poaceae
Sorghum lanceolatum	Poaceae
Sporobolus marginatus	Poaceae
Sporobolus festivus	Poaceae
Elusine indica	Poaceae
Sida cordifolia	Malvaceae: Malvoideae
Hypostis cancellata	Acanthaceae
Ipomoea aquatica	Convolvulaceae
Acanthospermum hispidum	Asteraceae
Hibiscus spp.	Malvaceae: Malvoideae
Amaranthus spp.	Amaranthaceae

mentioned in the flora of the desert ecological zone. Long drought spells after germination may cause complete disappearance of the plant from the rangeland. At present the plant has disappeared from many grazing areas of Butana. There is only one area from which seed stock can be collected for regeneration (reseeding) which is Jebel Elmundra in the Butana plain, where stands of under-grazed plants are found due to limited resources of livestock drinking water. Light clay soils are suitable for germination and establishment of uncovered mature seeds.

During an EIA survey for the proposed Kajbar hydropower dam, Belpharis was found in one of the seasonal water courses draining into the western bank of the River Nile. This could be the first record of the species in the desert zone of Sudan (**Figure 1** & **Figure 2**).

Figure 1. *Belpharis persica* in a dry seasonal wadi in Kajbar area (northern state).

Figure 2. Close up of *Blepharis persica*.

3.3. Provision of New Information on Endangered Species

Medemia argun (Martius) Wuttt. ex H. A. Wendi Family: Arecaceae (**Figures 3-5**).

Medemia is a monotypic genus endemic to the upper valley of the River Nile in Egypt and Sudan. Nowadays, there are only a few localities in which this palm is found, and in most cases the populations there consist of very few individuals. The Sudanese Nubian Desert east of the River Nile is the only region where the plant is now naturally growing. This region is so arid that consecutive years pass without any rain falling. Though the terrain is harsh with many rocky features, the Nubian Desert is characterized by a network of seasonal water courses that drain into and form the large Wadi-Gabgaba. These wadis could run with flash floods coming from seasonal rainfall in the Red Sea Hills. According to the International Union for the Conservation of Nature (IUCN) Red list Category and Criteria [8], the "Argoun Palm" is *critically endangered.*

Figure 3. Argoun palm showing a skirt of dry leaves.

Figure 4. Argoun palm tree with a cluster of fruits.

Figure 5. A dense population of Argoun palm.

During two visits to the eastern Nubian Desert in 2011 and 2015, several populations of Medemia have been encountered. The findings have manifested not only that Medemia still exists, but it inhabits many locations. Most striking is that the tree is in good condition and many fruiting trees and several young trees at different stages of growth have been encountered [9]. Nevertheless, the criterion upon which IUCN has categorized the Argoun palm as critically endangered—small and fragmented populations—is still valid.

3.4. Discovery of New Species (Nomennovum)

During a multidisciplinary expedition organized in 1979 by the Hydrobiological Research Unit (Later on amalgamated in the IES) of University of Khartoum to study the expected environmental and social impacts of the proposed Jonglei Canal in southern Sudan, a plant was observed growing in swamps at the joining of Bahr el Zeraf and a side lake. It was first taken to be a member of the Araceae family due to the sagitate shape of the leaf base. When several specimens were taken to Kew, London, it was discovered that the plant is not only a new species but a new genus of the poaceae (Graminae) family! It was aptly named *Suddia sagitifolia.* The discovery was considered as the declaration that *Suddia sagitifolia* is the only endemic aquatic plant in the Sudan (**Figure 6**) [10]. Endemic plants and animals are those that are unique to a specific geographic region. This makes them incredibly special and more vulnerable to extinction. Because they are only found in certain locations, they require special conservation efforts.

The plant was encountered by the author during a later field work mission in 2005 organized by the IES as part of an EIA study for Block 5B for the White Nile Petroleum Operating Company within Bahr el Zeraf Basin [2]. Very recently the same plant was found in some wetlands in Uganda and the plant was rated as a near endemic species. Nevertheless, the credit for this contribution of international dimension still remains to the EIA of seventies of the last century. Within the same region the smallest floating plant (*Wollfia hyaline*) in the former Sudan was also reported.

Figure 6. *Suddia sagitifolia* with bright green leaves in the former southern Sudan [2].

The available information on the distribution of flora on Sudan is old and limited. The EIA studies have availed valuable chances for the flora expert within the EIA team to visit many sites within Sudan. Some of these sites fall in regions which are otherwise inaccessible hadn't it been for the EIA missions. The lists are not exclusive yet useful for future monitoring and for providing a base to revise and update the flora list. The identification of the plants was not an easy task due to the paucity in taxonomy references as well as national taxonomists. So far the only available references are the three volumes of Andrews in the fifties of the last century [11] [12]. However, recently a significant contribution to alleviate this situation is the publishing of an annotated checklist on the plants of Sudan and South Sudan [13].

4. Conclusions

- Environmental Impact Assessment studies could be a valuable source of information in the field of biodiversity.
- The obtained lists could be a valuable baseline for future monitoring studies where changes due to natural and anthropogenic factors can be detected.
- The contribution could be many fold: in producing species list, in taxonomy update and in producing new information about species distribution.
- The studies revealed the paucity in endemic species in Sudan (only one).

Recommendations

- EIA reports should be used to document fauna as well.
- The documented lists should be used for future follow up and monitoring studies.
- EIA consulting companies need to be requested to adopt the most recent taxonomic nomenclature.

- More research should be directed towards endemic and endangered species in Sudan.

References

[1] UNEP (2002) Environmental Impact Assessment Training Resource Manual. (Eds) B. Sadler and M. McCabe.

[2] White Nile Petroleum Operating Company (WNPOC) (2005) ESIA Report for Block B, Southern Sudan.

[3] Ali, O.M.M. (2009) The Aquatic Plants of Sudan. In Dumont, H.,Ed., The Nile: Origin, Environments, Limnology and Human Use. Springer Publisher, Germany, 479-494. http://dx.doi.org/10.1007/978-1-4020-9726-3_23

[4] Fichtner-UKCC (2009) Environmental Impact Assessment for Kajbar Hydropower Project.

[5] StarOil Petroleum Operation Company (2014) Crude Oil Pipeline from South Annajama EPF to the Landing Terminal of GNPOC Export Pipeline at Um Adara.

[6] Harrison, M.N. and Jackson, J.K. (1958) Ecological Classification of the Vegetation of the Sudan. Forest Bull No.2. (New Sereis): 1-45. Khartoum.

[7] Hassan M.H. (1974) An Illustrated Guide to the Plants of Erkawit. Khartoum University Press, Khartoum.

[8] IUCN (2010) Red List of Threatened Species. Version 2014.3.

[9] Ali, O.M.M. (2016) The Argun Palm, (*Medemia argun*), in the *Eastern* Nubian Desert of Sudan. *PALMS*, **60**, 145-153.

[10] Renovate, S.A., Lock, J.M. and Denny, P. (1984) A Remarkable New Grass Genus for the Southern Sudan. *Kew Bulletin*, **39**, 455-461. http://dx.doi.org/10.2307/4108589

[11] Andrews, F.W. (1950, 1953) The Flowering Plants of the Anglo-Egyptian Sudan. V. 1 & II. T. Buncle & Co. Arbroath, Scotland.

[12] Andrews, F.W. (1956) The Flowering Plants of the Sudan. V. III. T. Buncle & Co. Arbroath, Scotland.

[13] Derbyshire, I., Kordofani, M., Farag, I., Candiga, R. and Pickering, H. (Eds.) (2015) The Plants of Sudan and South Sudan: An Annotated Checklist. Royal Botanic Gardens, Kew, United Kingdom.

Permissions

All chapters in this book were first published in JEP, by Scientific Research Publishing; hereby published with permission under the Creative Commons Attribution License or equivalent. Every chapter published in this book has been scrutinized by our experts. Their significance has been extensively debated. The topics covered herein carry significant findings which will fuel the growth of the discipline. They may even be implemented as practical applications or may be referred to as a beginning point for another development.

The contributors of this book come from diverse backgrounds, making this book a truly international effort. This book will bring forth new frontiers with its revolutionizing research information and detailed analysis of the nascent developments around the world.

We would like to thank all the contributing authors for lending their expertise to make the book truly unique. They have played a crucial role in the development of this book. Without their invaluable contributions this book wouldn't have been possible. They have made vital efforts to compile up to date information on the varied aspects of this subject to make this book a valuable addition to the collection of many professionals and students.

This book was conceptualized with the vision of imparting up-to-date information and advanced data in this field. To ensure the same, a matchless editorial board was set up. Every individual on the board went through rigorous rounds of assessment to prove their worth. After which they invested a large part of their time researching and compiling the most relevant data for our readers.

The editorial board has been involved in producing this book since its inception. They have spent rigorous hours researching and exploring the diverse topics which have resulted in the successful publishing of this book. They have passed on their knowledge of decades through this book. To expedite this challenging task, the publisher supported the team at every step. A small team of assistant editors was also appointed to further simplify the editing procedure and attain best results for the readers.

Apart from the editorial board, the designing team has also invested a significant amount of their time in understanding the subject and creating the most relevant covers. They scrutinized every image to scout for the most suitable representation of the subject and create an appropriate cover for the book.

The publishing team has been an ardent support to the editorial, designing and production team. Their endless efforts to recruit the best for this project, has resulted in the accomplishment of this book. They are a veteran in the field of academics and their pool of knowledge is as vast as their experience in printing. Their expertise and guidance has proved useful at every step. Their uncompromising quality standards have made this book an exceptional effort. Their encouragement from time to time has been an inspiration for everyone.

The publisher and the editorial board hope that this book will prove to be a valuable piece of knowledge for researchers, students, practitioners and scholars across the globe.

List of Contributors

Nehad M. Abd El Moneam and Nabila E. Abd El Maguid
Biochemistry and Zoology Department, Faculty of Science, Alexandria University, Alexandria, Egypt

Amany M. El-Sikaily and Marwa G. Zaki
Marine Pollution Research Lab., National Institute of Oceanography & Fisheries, Alexandria, Egypt

Mohamed A. Sheradah
Marine Biotechnology Lab., National Institute of Oceanography & Fisheries, Alexandria, Egypt

Sheri L. Herron
Ecosustain, LLC, Dripping Springs, TX, USA

Andrew N. Sharpley, Kristofor R. Brye and David M. Miller
Department of Crop, Soil, and Environmental Sciences, University of Arkansas, Fayetteville, AR, USA

Lahua Jin, Chen Zhao and Jiayi Wang
School of Environment, Jinan University, Guangzhou, China

Jing Shi, Naiming Zhang, Chenguang Zhang and Boxian Wei
College of Resource and Environment, Yunnan Agricultural University, Kunming, China
Soil Fertilization and Pollution Remediation Engineering Laboratory of Yunnan Province, Kunming, China

Dorrit H. Lowsen
U.S. Department of State, Beijing, China

George A. Conway
Division of Global Health Protection, Center for Global Health, US Centers for Disease Control and Prevention, Atlanta, GA, USA

Wilfred N. Igboama
Department of Physics, Federal University, Oye-Ekiti, Nigeria

Nicholas U. Ugwu
Department of Physics, University of Botswana, Gaborone, Botswana

Nisha Soni, Robin A. Christian and Namrata Jariwala
Civil Engineering Department, SVNIT, Surat, India

Bushra Al Wahaibi, Haider Al Lawati and Salma M. Z. Al Kindy
Department of Chemistry, College of Science, Sultan Qaboos University, Muscat, Sultanate of Oman

Rengaraj Selvaraj
Department of Chemistry, College of Science, Sultan Qaboos University, Muscat, Sultanate of Oman
Laboratory of Green Chemistry, LUT Savo Sustainable Technology, Lappeenranta University of Technology, Mikkeli, Finland

Mika Sillanpää
Laboratory of Green Chemistry, LUT Savo Sustainable Technology, Lappeenranta University of Technology, Mikkeli, Finland

Kezhen Qi
Institute of Catalysis for Energy and Environment, College of Chemistry and Chemical Engineering, Shenyang Normal University, Shenyang, China

Sathish Babu Soundra Pandian
CAARU, College of Science, Sultan Qaboos University, Muscat, Sultanate of Oman

Mohammed A. Meetani
Department of Chemistry, United Arab Emirates University, Al Ain, UAE

Younghun Kim
Department of Chemical Engineering, Kwangwoon University, Seoul, South Korea

Thomas Alsmo
Library Research, Olten, Switzerland

Handojo Djati Utomo, Keng Xuan Donovan Tan, Zhi Yi Daniel Choong, Jia Jia Yu and Jie Jun Ong
Division of Civil Engineering, School of Architecture and the Built Environment, Singapore Polytechnic, Singapore, Singapore

Zheng Bang Lim
Advanced Material Technology Centre, Technology Development Office, Singapore Polytechnic, Singapore, Singapore

Wedad Rayif Alharbi
Department of Physics, Faculty Science, King Abdulaziz University, Jeddah, Saudi Arabia

Richard C. Smardon
Department of Environmental Studies, SUNY College of Environmental Science and Forestry, Syracuse, USA

S. M. Kerstens
Royal HaskoningDHV, Amersfoort, The Netherlands

G. Hutton
UNICEF, New York, USA

I. Firmansyah and G. Zeeman
Sub-Department of Environmental Technology, Wageningen University, Wageningen, The Netherlands

I. Leusbrock
AEE INTEC, Gleisdorf, Austria

Osman M. M. Ali, Mohamed F. Idris
Institute of Environmental Studies, Khartoum, Sudan

Index